Springer Transactions in Civil and Environmental Engineering

Springer Transactions in Civil and Environmental Engineering (STICEE) publishes the latest developments in Civil and Environmental Engineering. The intent is to cover all the main branches of Civil and Environmental Engineering, both theoretical and applied, including, but not limited to: Structural Mechanics, Steel Structures, Concrete Structures, Reinforced Cement Concrete, Civil Engineering Materials, Soil Mechanics, Ground Improvement, Geotechnical Engineering, Foundation Engineering, Earthquake Engineering, Structural Health and Monitoring, Water Resources Engineering, Engineering Hydrology, Solid Waste Engineering, Environmental Engineering, Wastewater Management, Transportation Engineering, Sustainable Civil Infrastructure, Fluid Mechanics, Pavement Engineering, Soil Dynamics, Rock Mechanics, Timber Engineering, Hazardous Waste Disposal Instrumentation and Monitoring, Construction Management, Civil Engineering Construction, Surveying and GIS Strength of Materials (Mechanics of Materials), Environmental Geotechnics, Concrete Engineering, Timber Structures.

Within the scopes of the series are monographs, professional books, graduate and undergraduate textbooks, edited volumes and handbooks devoted to the above subject areas.

More information about this series at http://www.springer.com/series/13593

Gaurav Srivastava · Pravinray D. Gandhi

Performance of Combustible Façade Systems Used in Green Building Technologies Under Fire

Springer

Gaurav Srivastava
Department of Civil Engineering
Indian Institute of Technology Gandhinagar
Gandhinagar, Gujarat, India

Pravinray D. Gandhi
Underwriters Laboratories Inc.
Chicago, IL, USA

ISSN 2363-7633 ISSN 2363-7641 (electronic)
Springer Transactions in Civil and Environmental Engineering
ISBN 978-981-16-3111-5 ISBN 978-981-16-3112-2 (eBook)
https://doi.org/10.1007/978-981-16-3112-2

This Springer imprint is published by the registered company Springer Nature Singapore Pte Ltd.
The registered company address is: 152 Beach Road, #21-01/04 Gateway East, Singapore 189721, Singapore

Foreword

I spent most of my professional career at the Indian Institute of Technology Kanpur working in the area of earthquake engineering. My journey with the Indian Institute of Technology Gandhinagar (IITGN) as its first Director started in 2009, about a year after its establishment. My first introduction to Underwriters Laboratories (UL) was in 2007 when I was invited to the Fire Safety Council organized by them at Delhi. The relationship continued beyond that meeting. In 2009, before taking the office as the Director of IITGN, I was flying from China to Kanpur via Delhi. During the layover at Delhi, there was an opportunity to meet Dr. Pravinray D. Gandhi, Mr. R. A. Venkitachalam, Mr. V. Jagdish, and others of UL. In that meeting, I offered to UL a major initiative focused on public safety at IITGN—a newly founded institute with almost no faculty and laboratory facilities, but with promising undergraduate students. UL agreed to this proposition, and this is how foundations of the 'Centre for Safety Engineering' at IITGN were laid. The focus of this Centre was to improve safety across different walks of life through research and development activities and to disseminate the knowledge generated. We undertook several student challenges and projects in the areas of water quality, kitchen fire safety, electrical fire safety, etc. In due course of time, fire safety became one of the most vibrant areas of work for the Centre.

The project on façade fire safety was initiated in June 2016 when Mr. R. A. Venkitachalam (then Vice President, Safety, UL India) proposed the idea of development of one of a kind fire testing facility—a three-story full-scale structure—at IITGN for assessment of different façade systems subjected to fire. From UL's side, the project was being led by Mr. V. Jagdish (then Head Regulatory, UL India) with R&D support being provided by Dr. Pravinray D. Gandhi. From IITGN's side, Dr. Gaurav Srivastava from Civil Engineering and Dr. Chinmay Ghoroi from Chemical Engineering (both involved with the Centre for Safety Engineering) joined hands with UL to take up this challenging project. The project was challenging on two fronts: First, it had no precedence which meant the design and construction aspects were unique, and second, it had a tight timeline, wherein the discussions started in June 2016 and the first set of experiments were to be performed in December 2016.

The project team worked in close coordination, and the first full-scale fire experiment (possibly the first of its kind in India) was performed on December 8, 2016.

Since then, a number of experiments have been performed at this facility with different industry partners. The success of this facility lies in the fact that multiple industry players have collaborated with IITGN to carry out full-scale fire experiments of different kinds—all of them with a single aim—to enhance fire safety of modern buildings.

Dr. Gaurav Srivastava and Dr. Pravinray D. Gandhi have been closely engaged with all of the experiments performed at this facility and have helped gain significant insights into the behavior of combustible and non-combustible façade systems subjected to fire. This book, written by them, provides an in-depth view and analyses of seven full-scale fire experiments performed at the fire testing facility at IITGN. This provides a wealth of experimental data, which is usually difficult to find—real fire accidents do not yield any data beyond occasional video footage (since buildings are not instrumented, in general), and laboratory fire experiments do not simulate a real building fire (due to different control mechanisms and being of smaller scales). The discussions include valuable photographs from different stages of experiments correlated with temperature data and other insights. Further, data from bench-scale experiments of the materials used in full-scale experiments are also presented—which can be very useful for those interested in computer simulation of façade fires.

Furthermore, discussions are presented in the context of existing test methods and relevant theoretical background. Rationale behind some of the standard test methods adopted by different countries is presented with relevant scientific background in a fundamental sense. When this is considered with the experimental case studies, it helps in appreciating the strength and shortfalls of different methods vis-à-vis real fire experiments. A number of practical aspects involved in the development of an experimental facility of this scale and conducting such experiments are discussed in detail. This can be of great help to those who wish to create similar facilities to enable studies to enhance fire safety of modern infrastructure.

Overall, this book is the outcome of a long and strong partnership between two institutions—IITGN and UL—interested in enhancing public safety. Dr. Gaurav Srivastava and Dr. Pravinray D. Gandhi present hard to get data for researchers, practical insights for field engineers, and theoretical background for students who may be interested in learning more about the behavior of façade systems in fire.

Sudhir Kumar Jain
Professor of Civil Engineering and Director
Indian Institute of Technology Gandhinagar
Gandhinagar, India

Preface

The idea of this book was conceived during the execution of a series of full-scale fire experiments being conducted at IIT Gandhinagar (IITGN) in collaboration with Underwriters Laboratories (UL) to study the behavior of combustible façade systems under real fire scenarios. The project was initiated in 2016 when Mr. V. Jagdish (then Head Regulatory, UL India) began discussions on the need of full-scale testing of façade systems to validate their performance in real fire scenarios. The experimental facility was planned and constructed within a short span of three months and happens to be the only of its kind facility in India. It has facilitated testing of a variety of different façade systems and has generated a wealth of experimental data on real fire performance of different façade (and building) components. After a number of tests were conducted, the authors felt the need for a thorough documentation of the procedures and a detailed scientific exposition of the observations to enable improvements in the current state of the art and state of the practice.

The uniqueness of the experiments discussed in this book is their being close to a realistic building fire. The test facility is a three-story full-scale structure where realistic fire scenarios were developed with real furniture (as fuel) and installations of different systems. Fires were not controlled in any way after ignition and were allowed to grow naturally. Extensive instrumentation in the room of fire origin and one above it enabled a deeper understanding of fire growth starting with room contents and then growing to involve the building façade and subsequent fire growth to upper floors. While a number of laboratory-scale experiments are available to quantify fire behavior of such systems, they usually entail installation of a sub-system and employ controlled fire scenarios. This limits the usability of the data in extrapolating the performance of tested systems in real scenarios. On the other hand, real building fire accidents subject all the components to a fire as a system and post-fire analyses are based on limited data (usually video footage/photographs from different sources). The test facility and the experiments presented in this book bridge this gap—they utilize real building system installation and fire scenarios and provide measured data of various kinds.

The book can be utilized at many levels by different stakeholders. Researchers can make use of the experimental data and test configurations at different scales (component/material-level experiments and full-scale experiments) to develop computer models for façade fires—an active and challenging area of work. Architects can consider the real fire performance of different building components and features—size, shape, geometry, materials, fire protection, firestop, façade, etc.—to make a more informed choice for designs. Manufacturers can relate the performance of different systems (façade materials, fire protection system, firestop, sprinklers, etc.) discussed in the book with standardized certification tests that they might be doing. Results in such experimental settings can provide useful validation of designs and confidence for clients. Practicing engineers can treat this book as a guide for construction, installation, and performance of different building components. Many practical aspects of installations (construction details, workmanship issues, etc.) have been discussed.

Students should find this book to be a useful resource in learning about the fire behavior of façade systems. Theoretical background of relevant heat transfer phenomena, fire plume behavior, effects of window size on heat flux, flame height, and heat release rate has been presented and related to existing test methods and, also, real fire performance of different façade systems. Guidance on use of computer simulation tools such as appropriateness of grid size and its efficacy in assessing different scenarios is also provided. Some key length scales associated with window fires are discussed and illustrated in detail.

Gandhinagar, India Gaurav Srivastava
Chicago, USA Pravinray D. Gandhi

Acknowledgements

Experiments of the scale presented in this book do not become possible without active involvement of a large number of stakeholders. Detailed documentation and coherent presentation of the inferences in the form of a book also require support from colleagues and family.

Acknowledgments are due to IIT Gandhinagar (IITGN) and Underwriters Laboratories (UL) for providing financial and logistic support for making the project viable. A special mention of Mr. V. Jagdish (then Head Regulatory, UL India) is necessary, for it was his passion that provided the initial momentum to the series of experiments that were performed. Other team members from UL—Dr Stephen Kerber (Vice President, UL-FSRI), Mr. R. A. Venkitachalam (Vice President, UL), Ms. Monalisa Das (Project Engineer, BLST), Mr. Aravind Chakravarthy (Building Inspection), Mr. Karthikeyan G. (Building Inspection), and Mr. Abhishek Kandoi (Senior Project Engineer)—played an active role in making this project a success.

Valuable support from a team of IITGN is also acknowledged—Prof. Sudhir K. Jain (Director), for his enthusiastic support and encouragement for the project; Prof. Chinmay Ghoroi (Coordinator, Centre for Safety Engineering) for his active involvement; Prof. S. P. Mehrotra and Prof. Amit Prashant (Dean, Research and Development) for their support; Prof. Harish P. M. (Dean, Campus Development) for his involvement during many situations of crisis; Mr. G. C. Chaudhary and Mr. V. K. Baghel (Superintending Engineer) for their leadership in resolving site and construction-related issues; and Mr. Laxmikant Mishra and Mr. Ravi Soni (Assistant Engineer) and Mr. Dharmendra Panchal (Junior Engineer) for their prompt support for various site-related requirements.

A large number of students participated actively in helping with different aspects during construction of the facility and experiments. The largest share of acknowledgments is due to Mr. Dharmit Nakrani (currently working with Building Research Establishment, UK), who took this work up as his doctoral thesis work and worked tirelessly throughout the process; he worked on the ground with contractors, helped make drawings and sketches, and carried out data extraction, analysis, and interpretation of various experiments. Support from many other students of the fire group at IITGN is acknowledged—Dr. Ravi Prakash, Mr. Nasar A. Khan, Mr. Harshal

Lambhate, Mr. Harshit Nema, Mr. Sarana H. Kota, and Ms. Laxmi Tiwari. Commitment of support staff—Mr. Pradip Ninama, Mr. Mahendra Bhai, and Mr. Dhanabhai Rabari—was a great asset that enabled successful completion of the experiments. The help provided by Mr. Ben Gaudet with respect to the computational analyses required for this book is gratefully acknowledged.

Acknowledgment is due to Mr. Kumudchandra Suthar—the main contractor—who undertook the construction of this unique facility and finished it in a record time of three months. He accommodated of many late changes to different systems, which were required as such a facility was being constructed for the first time and the design was constantly evolving.

The support from Mr. Mukesh Shah (Shah Bhogilal Jethalal & Bros.) and Mr. Abhay Purandare (Fire Consultant) is gratefully acknowledged. Support from different industry partners for different experiments—Hilti India Pvt Ltd., Specified Technologies Inc., The Lubrizol Corporation, and Shah Bhogilal Jethalal & Bros.—is acknowledged.

In projects and writings of this scale, the sacrifice of family members usually remains in the background. Continuous engagement away from the family (during execution of experiments, data extraction and analyses, and writing) made everything possible. Gaurav wishes to acknowledge selfless support from his wife Akanksha and son Ishan that allowed him to carry out this project and develop this manuscript. He is also indebted for the blessings of his parents, Pushpa and Ashok, and motivation from his sister, Jaya.

Contents

About the Authors

Dr. Gaurav Srivastava is an Associate Professor of Civil Engineering at Indian Institute of Technology Gandhinagar (IITGN). His research interests lie in assessment of structural and non-structural building components subjected to fire. He has led several studies in this area. While some of these studies aimed at fundamental research and development, others dealt with contemporary industrial issues. Prior to joining IIT Gandhinagar, he received his doctoral degree in Civil Engineering from the University of Minnesota and worked at the University of Notre Dame.

Dr. Pravinray D. Gandhi is Director of Research at Underwriters Laboratories Inc. with focus on fire safety. He has worked on research on a number of areas including smoke detection, fire growth and fire suppression. He has also contributed to the development of new fire performance test standards, fire hazard assessment methods, and has served on ASTM, NFPA, and IEC standards bodies as a subject matter expert.

Abbreviations

AC	Alternating Current
ACP	Aluminum Composite Panel
CCD	Charge-Coupled Device
CFD	Computational Fluid Dynamics
CPVC	Chlorinated Polyvinyl Chloride
DAQ	Data Acquisition System
DC	Direct Current
DGU	Double Glazed Unit
DVR	Digital Video Recorder
EIFS	Exterior Insulation and Finish System
ENR	Expanded Nitrile Rubber
EPS	Expanded Polystyrene
ETICS	External Thermal Insulation Cladding System
FDS	Fire Dynamics Simulator
FF	First Floor
GF	Ground Floor
HPL	High Pressure Laminate
HRR	Heat Release Rate
HRRPUV	Heat Release Rate Per Unit Volume
HVAC	Heat Ventilation and Air Conditioning
LVDT	Linear Variable Differential Transformer
MDF	Medium Density Fiberboard
MS	Mild Steel
NRC	National Research Council (Canada)
PE	Polyethylene
PIR	Polyisocyanurate
POP	Plaster of Paris
PU	Polyurethane
PUR	Rigid Polyurethane
PVC	Polyvinyl Chloride
PVDF	Polyvinylidene Difluoride
QGU	Quadruple Glazed Unit

SF	Second Floor
SGU	Single Glazed Unit
SPI	Society of Plastics Industry
TGU	Triple Glazed Unit
UBC	Uniform Building Code
UPVC	Unplasticized Polyvinyl Chloride
XPS	Extruded Polystyrene

Symbols

A_V	Area of window (m^2)
A_T	Total enclosure surface area (m^2)
C	Specific heat (J/kgK)
C_g	Specific heat of gas (J/kgK)
$C_{p,\infty}$	Ambient specific heat at constant pressure (J/kgK)
D	Characteristic dimension/length scale (m)
D_h	Window hydraulic diameter (m)
$\overset{*}{D}$	Non-dimensional length scale (unit-less)
F_V	Opening/ventilation factor ($m^{0.5}$)
g	Acceleration due to gravity (m/s^2)
h_c	Convective heat transfer coefficient (W/m^2K)
h_r	Radiative heat transfer coefficient (W/m^2K)
H_V	Height of window (m)
ΔH_{OX}	Heat released per unit mass of oxygen (J/kg)
λ	Thermal conductivity (W/mK)
l_1, l_2	Length scale (m)
$\overset{*}{l_1}, \overset{*}{l_2}$	Non-dimensional length scale (unit-less)
L	Length/thickness (m)
$\dot{m}_g$	Mass flow rate of gas (kg/s)
$\dot{m}_{in}$	Air inflow rate (kg/s)
n	Window aspect ratio (unit-less)
$\dot{q}''_s$	Heat flux normal to surface (W/m^2)
$\dot{q}''_{cond}$	Conductive heat flux (W/m^2)
$\dot{q}''_c$	Convective heat flux (W/m^2)
$\dot{q}''_r$	Radiative heat flux (W/m^2)
$\dot{q}''_t$	Total heat flux (W/m^2)
Q	Heat release rate (W)
$\overset{*}{Q}$	Non-dimensional heat release rate (unit-less)
Q_c	Convective heat release rate (W)
Q_{fl}	Heat release rate at flashover (W)
Q_{ex}	Excess heat release rate (W)

Q_{in}	Input heat release rate (W)
Q_{st}	Heat release rate constrained by oxygen (W)
Q_t	Total heat release rate (W)
ρ	Density (kg/m^3)
ρ_g	Gas density (kg/m^3)
ρ_∞	Ambient density (kg/m^3)
r_0	Length scale (m)
R	Thermal resistance (m^{2}K/W)
θ	Non-dimensional temperature (unit-less)
t	Time (s)
t_{burn}	Time of burning (s)
T	Temperature (K)
T_s	Surface temperature (K)
T_g	Gas temperature (K)
T_∞	Ambient/far-field temperature (K)
W_v	Width of window (m)
x	Spatial coordinate (m)
Y_{ox}	Mass fraction of oxygen in air (unit-less)
z	Spatial coordinate (m)
z_0	Location of virtual origin of fire (m)
z_f	Height of flame (m)
z_n	Height of neutral pressure plane (m)

Chapter 1
Introduction

Green buildings aim to increase the efficiency of use of energy, water and materials; lower the impact on environment during their construction and operations; and reduce building impact on human health. Several organizations across the globe—World Green Building Council (founded 2002); US Green Building Council (founded 1993)—actively engage in discussions and initiatives to improve greenness of built infrastructure. The US Green Building Council developed the Leadership in Energy and Environmental Design (LEED) certification program which evaluates built-environment projects on key areas of human and environmental health: sustainable site development, water savings, energy efficiency, materials selection, and indoor environmental quality. Many countries have developed such mechanisms, e.g. Deutsche Gesellschaft für nachhaltiges Bauen e.V. (DGNB) in Germany, Miljöbyggnad in Sweden, Building Research Establishment Environmental Assessment Method (BREEAM) in the UK, BCA Green Mark Scheme in Singapore, BEAM in Hong Kong, CASBEE in Japan, Green Star SA in South Africa, Pearl Rating System for Estidama in UAE, and Green Rating for Integrated Habitat Assessment (GRIHA) in India.

Building façades plays an important role in the green strategies of a building by providing weather and thermal insulation, natural light and ventilation in addition to enhancing aesthetics of the building and providing safety and comfort to building occupants. Herzog (2018) provided a list of factors that are considered for optimizing façade performance as shown in Table 1.1.

Historically, building façades utilized predominantly non-combustible materials such as brick masonry, precast concrete panels, steel framing, stone cladding, etc.. Use of combustible materials in façades started in the mid-1970s in the wake of an energy crisis and rising petroleum prices with a motivation to improve thermal efficiency of buildings. This involved use of plastic-based thermal insulation materi-

G. Srivastava and P. D. Gandhi, *Performance of Combustible Façade Systems Used in Green Building Technologies Under Fire*, Springer Transactions in Civil and Environmental Engineering, https://doi.org/10.1007/978-981-16-3112-2_1

Table 1.1 Relative importance of façade functions

Parameter	Optimization	Purpose
Aesthetics	Maximize	Improves building attractiveness.
Flammability	Minimize	Reduces contribution of façade to flame spread and increases occupant safety.
Moisture Ingress	Minimize	Reduces moisture in façade to mitigate mold and dampness in the building.
Weight	Minimize	Reduces weight to reduce construction time and cost of structural support for installing façade.
Thickness	Minimize	Increases available living area inside the building.
Structural Stability	Maximize	Increases the façade's ability to withstand self-weight, wind, and other live loads.
Thermal Insulation	Maximize	Improves building energy efficiency and comfort to building occupants.
Lighting	Maximize	Improves building energy efficiency and comfort for building's occupants. Enhances human experience.
Sound insulation	Maximize	Block external noise to increase comfort for building occupants.
Return on Investment	Maximize	Improves building economics.

als. With this change came the recognition that these materials can get involved and contribute to a fire started externally or a room and contents fire that progresses to the outside of the building.

Participation of façade systems in intensifying building fires in recent years has been a point of concern across the globe. Bonner (2020) created a list of high-rise fire accidents in which façade systems actively participated in a fire. A graphical representation of the number of façade fires since 1990s is shown in Fig. 1.1. A rather uncomfortable rising trend in façade fires is evident from this data. What is of even greater concern is that this data includes fires of high-rise buildings only. Such buildings are expected to conform to the prevalent codes and standards and are usually better engineered compared to a large number of other buildings, especially in developing countries.

More insights can be gained by looking at the same data country-wise, as shown in Fig. 1.2. This bar chart includes countries where more than one façade fire has

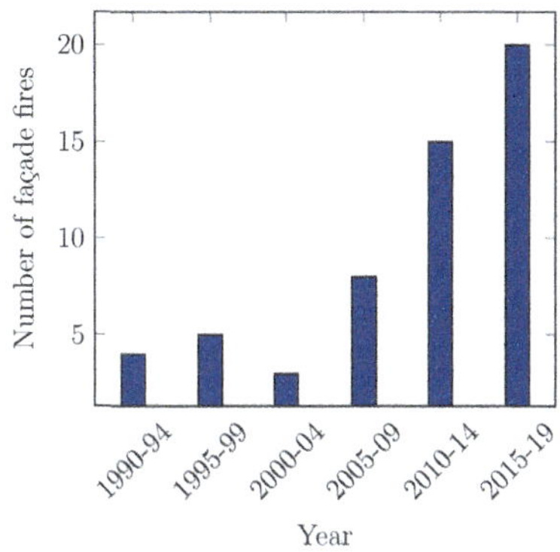

Fig. 1.1 Rise in the number of high-rise façade fires in recent years (Bonner, 2020)

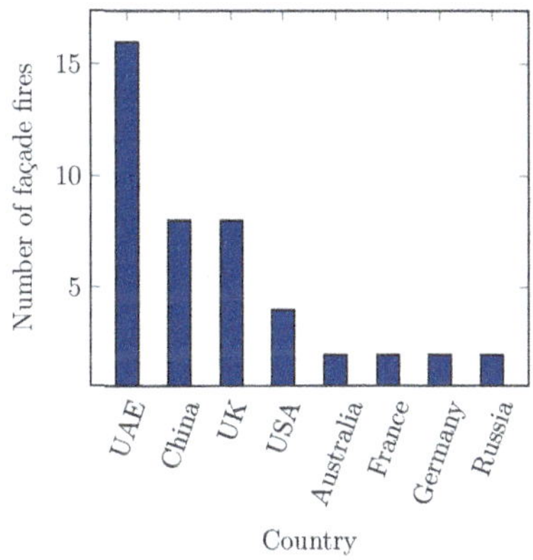

Fig. 1.2 Recent high-rise façade fires by the country

taken place since 1990 as per the data of Bonner (2020). UAE tops the list with many notable façade fires (e.g. fires at The Address Hotel, The Torch). China and UK are next with eight façade fires each followed by USA with four fires. The most recent major façade fire accident that stirred the engineering community and initiated serious discussions on façade systems and materials occurred at Grenfell Tower, London, UK. See Grenfell (2019) for a detailed account of various discussions and investigations that have happened after the Grenfell Tower fire.

Almost all of the façade fires documented by Bonner (2020) can be attributed to the use of combustible materials in façade. There have been other notable fires where internal spread of hot gases/smoke within a building due to inadequate firestop systems caused major losses. These include the First Interstate Tower fire in Los Angeles (1988) and the ESIC Hospital fire in Mumbai, India (2018). Improper firestop systems can pose a threat to life safety much before the fire grows in size. In combustible façades, providing adequate firestop is challenging due to the additional heat exposure and hot movements caused by the burning façade components (mainly panels).

Different countries have developed codes and standards to enable testing and characterization of the fire performance of façade systems. White (2015) provided a detailed overview of existing standard test methods of different scales (full, intermediate, and small) along with presenting fire incident case studies of several façade fires. Their discourse resulted from a project aimed to characterize fire statistics, incidents, and review of existing test methods with the objective to create recommendations for further work to improve the existing state of affairs. Currently available testing methods cover a wide range of scenarios and conditions; test duration and acceptance criteria vary as well. There seem to be a need to consider different design requirements of façade systems and develop a harmonized approach toward their fire safety assessment which is informed by realistic fire scenarios.

Chapter 2 discusses key design parameters considered for façades concerning the performance requirements mentioned in Table 1.1. A discussion on the use of materials for thermal insulation and their fire behavior is presented along with the key heat transfer fundamentals. Use of different materials for thermal insulation and their potential for enhancing fire hazards is also discussed.

Chapter 3 presents a discussion on different materials used in different components of a façade in the context of their fire behavior. It also presents relevant test data of bench-scale experiments performed on combustible components of façade that were used in full-scale experiments presented in Chap. 5.

Chapter 4 presents a detailed discussion on heat transfer, flame heights, and other characteristics of window fires along with a discussion related to certain standard façade testing methods at different scales. The rationale behind the genesis of different test methods is discussed vis-a-vis observations from different experiments. A discussion on the necessity and use of computer simulations in improving the understanding the thermal exposure from façade fires is also presented. While computer simulation of façade fires is exceedingly complex due to several coupled physical phenomena, some recent studies have paved way toward façade fire simulations and demonstrated the importance of such simulations.

Chapter 5 presents detailed case studies of seven full-scale façade fire experiments that have been carried out at IIT Gandhinagar in collaboration with Underwriters Laboratories. First, the design of the experimental facility is detailed along with construction, instrumentation, and fire protection. Subsequently, full-scale experiments are discussed in detail along with timelines correlated with photographs and time–temperature data at various stages. Behavior and failure mechanisms of different components, such as cladding frame, façade panels, connection mechanisms, passive fire protection, and firestop systems, are discussed.

Finally, the main conclusions and future directions are presented in Chap. 6. The authors hope that discussions presented here will help in understanding the key performance metrics for design of façade systems. The case studies presented here provide insights in façade fires before, during and after the fire has taken place. Such insights are oftentimes difficult to obtain through a post-fire study of a real building fire as one needs to rely on eyewitness accounts, unavailability of suitable camera angles (in case videos are available at all), and unavailability of temperature data. This also highlights the usefulness of such a full-scale experimental facility which can replicate real building fires but has an advantage of being optimally instrumented to provide the relevant data.

Chapter 2
Aspects of Design of Façade Systems

Herzog (2004), in his 'façade construction manual,' provides an excellent perspective on the various design considerations and materials for façade systems. Many other good documents on design of façade systems are available. For reference, see Donaldson (1991)—Exterior wall systems: glass and concrete technology, design, and construction, Merritt (2001)—Building design and construction handbook, Heywood (2021)—Best practice for the specification and installation of metal cladding and secondary steelwork, Aksamija (2013)—Sustainable facades: design methods for high-performance building envelops, and Abrantes (2017)—The prefabrication of building facades. The objective of this chapter is to discuss and introduce certain essential features of façade design pertinent to the current context, and not to present in-depth design methodology, which is available in the aforementioned references.

2.1 Classification of Façade Systems

Façade systems can be classified in different ways depending on their attributes. This section presents some of the key classifications of façade systems.

2.1.1 Structural and Non-structural Façade Systems

A façade system designed as a load-bearing component of a building classifies as a structural façade. Many buildings designed with load-bearing masonry walls or reinforced concrete walls fall under this category. The external surfaces may be exposed or plastered; within plastering also, a variety of finishes can be used, e.g., smooth, textured, grit, etc. Since the load-bearing requirement invariably mandates

G. Srivastava and P. D. Gandhi, *Performance of Combustible Façade Systems Used in Green Building Technologies Under Fire*, Springer Transactions in Civil and Environmental Engineering, https://doi.org/10.1007/978-981-16-3112-2_2

the use of traditional building materials (masonry or concrete), such façades usually incorporate windows and other openings to establish connections of the inside of the building with the outside. Structural façades are not very common in modern buildings primarily due to the fact that use of load-bearing walls limits the height of the building; high-rise buildings typically involve more efficient load-bearing systems (e.g., moment resisting frames), wherein the role of walls is limited to provide partitions within the building (in tall buildings, shear walls are also used).

In such buildings, the external envelope is not required to carry any structural load. This allows more creative designs and choice of materials for the façade. Such façades are classified as non-structural façades. While masonry and concrete can still be used in such scenarios, the low load demand allows use of other materials such as glass (the most used modern material for façades) and metal composite sheets. Non-structural façades, also called curtain walls (akin to partition walls used inside the building), are typically designed to carry their own weight, wind pressure, and seismic forces deemed to be transmitted from the main structural system of the building. Stone cladding is also a widely used façade system (Lewis 1995).

2.1.2 Single-, Double-, Multi-pane Glazing

When using glass as the glazing unit for curtain walls, a Single Glazed Unit (SGU) implies the use of a single pane of glass in each panel of the façade system. Such type of façade systems is usually inefficient in providing climate control inside the building and is hence not used in major projects. In a Double Glazed Unit (DGU), a single façade panel comprises two glass panes separated by a gap (filled with air or vacuum). Due to the gap, DGUs provide better thermal insulation of the building and hence better climate control. The provision of two surfaces also allows use of different surface coatings (e.g., a solar reflective coating on the outer glass), better collision protection (breaking of one glass does not expose the building to the environment), and improved fire protection (SGU usually breaks earlier than DGU).

Along similar lines, multi-pane glazing systems are employed. A Triple Glazed Unit (TGU), comprising of three glass panes (and two cavities) allows more control than DGU. Many buildings in Nordic countries employ TGUs. Likewise, a Quadruple Glazed Unit (QGU) can be used to provide even more control over the building climate. Commercially, glazings up to six panes are available. Of course, the intricacies of manufacturing, demand on detailing and cost increase with the number of panes being used.

2.1.3 Insulated and Ventilated Façade Systems

An insulated façade system, typically called Exterior Insulation and Finish System (EIFS) or External Thermal Insulation Cladding System (ETICS) entails the use of thermally insulating materials on the exterior walls of a building to provide isolation from the climatic conditions. The thermal insulation is typically finished with a water resistant material. Sometimes, a detailed drainage system is provided within the EIFS to ensure proper functioning. Ventilated façade system utilizes an air cavity between two surfaces (e.g. between the building wall and the external cladding or between two glazing units). A natural (sometimes forced) convective air flow is established within this cavity, which enables cooling of the building as hot air escapes from top of the cavity while cool air ingresses from the base. A combination of insulation boards and ventilated system is also common (e.g. the Grenfell Tower, London utilized Polyisocyanurate (PIR) boards as the thermal insulation and ACP as the external façade with an air gap between the two).

2.1.4 Combustible and Non-combustible Façade

While the aforementioned classifications were based on functional requirements of the façade systems, primarily in terms of their thermal insulation capability, the classification of combustible vs. non-combustible essentially arises from these functional classifications. Combustibility of a façade depends on the types of materials used in its construction. Interaction between the various components is equally important. While a traditional façade comprising solely of non-combustible materials (e.g. masonry or precast concrete panels) poses no risk of fire propagation, most modern façade systems utilize a combination of combustible and non-combustible materials. Non-combustible materials include glass (for panels) and steel (for cladding frame) while combustible materials include ACP and High Pressure Laminate (HPL) (for panels), and Polyvinyl Chloride (PVC) and aluminum (for cladding frame). It has been found in many studies that the spandrel area plays an important role in the overall fire safety strategy of a building with a combustible façade. Details on the existing code based provisions and performance of combustible façade systems in real building fires will be discussed in subsequent chapters.

2.1.5 Stick, Semi-unitized, and Unitized Systems

Façade systems can be classified based on the method of installation such as conventional stick type, semi-unitized type, and unitized type. Stick type installations involve on-site installation of transoms and mullions to the building columns/slabs through metal brackets. Subsequently, cut to size glass (or other façade panel) is fixed on to the transoms/mullions through pressure plates or mechanical fasteners. In this type of construction, about 90% of the work is done at site and is usually the cheapest to execute among the three different methods. Since it requires elaborate scaffolding outside the building for installation, it is suitable mainly for low-rise buildings.

Unitized systems are the other end of the construction spectrum where only about 10% of the work is carried out at site—this is installation of brackets anchored to the building columns/slabs. Pre-framed façade panels of full story height are manufactured in factory and are brought to site and installed on these brackets. Since most of the work is done at a factory, unitized systems offer better quality and workmanship but are more expensive to execute. They are particularly suited when the desired level of elevation aesthetic is high, buildings are tall, or the area to be covered with façade is large.

Semi-unitized systems are midway between the stick and the unitized systems. Transoms and mullions are installed on-site just as in the stick method. Cut to size glass (or other façade panel) is structurally glazed to aluminum sub-frames by structural sealant, in a factory. The sub-frame is then installed on the grid work of transoms and mullions. About 50% work is done at the site. They offer somewhat better quality control and appearance compared to stick type installation.

2.2 Types and Components of a Façade System

A typical modern façade system has different components, as shown in Fig. 2.1. A cladding frame (typically made of aluminum) is secured to the floor slab through mechanical fasteners. The connection of the frame to the slab usually entails a gap of 100–250 mm. Façade panels (glass, ACP, MDF, etc.) are secured to the cladding frame either through pressure tapes or via mechanical fasteners, depending on the design requirements. Gaps between the panels and the frame are to be filled with appropriate sealing materials. In the spandrel area, good designs provide for a fire insulation. The gap between the slab and the façade is also filled with appropriate sealing systems, called firestop systems.

2.3 Functional Requirements of Façades

The primary functional requirement of façade systems is to provide an 'envelope' to the building in order to provide a certain isolation to the inside of the building from the outside environment. This enables lowering the overall electricity costs (by reducing the demand on air conditioning), thereby improving the living conditions. From the inside of a building, its façade can be viewed as a 'large' window, whereas from the outside, a façade aims to provide a characteristic appearance to the building which distinguishes one building from the other. The key requirements for a façade system to fulfill these functional demands are discussed in this section.

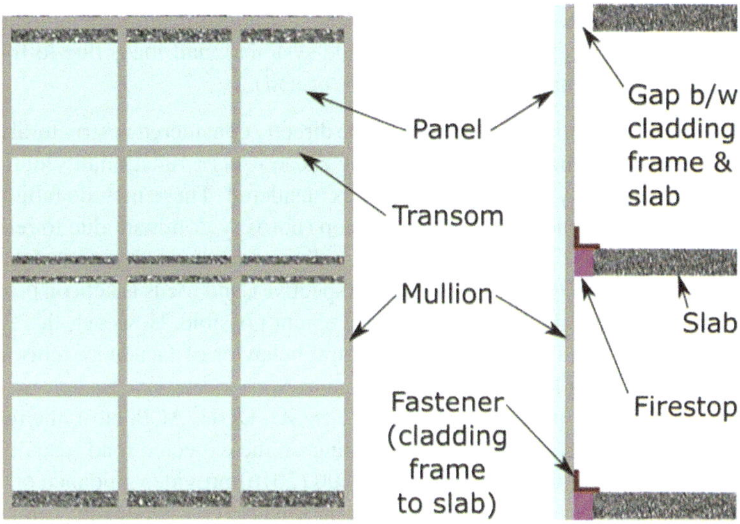

Fig. 2.1 Components of a façade system shown in front view (left) and side section elevation (right)

2.3.1 Structural Safety

From a structural safety point of view, façade systems need to be designed for the following loads/actions:

- Fixed/dead loads: These are primarily gravity loads originating due to self-weight of façade components (mullions, transoms, and panels).
- Imposed/live loads: Usually there are no imposed loads on façade systems unless some special provisions are made (e.g., safety hooks for cleaning).
- Wind loads: These are the dominant loads that act on façade systems. Both negative and positive wind pressures are required to be considered, i.e., low wind pressure inside building with high wind pressure outside creating a net inward force, and high wind pressure inside building with low wind pressure outside creating a net outward force. In areas where wind gusts are expected (e.g., coastal areas), special forces considering cyclonic winds, etc. also are required to be appropriately included in the analysis and design. In such cases, a dynamic analysis may be required.
- Earthquake loads: Earthquake loads will invariably be transferred to the façade systems indirectly through their fixing mechanisms to the main building and can be considered depending on the requirement. However, non-structural façade systems, usually much lighter than the building, may not require inclusion of earthquake forces unless they are being designed for high functional demand (e.g., glass panels should not break during an earthquake of a given intensity).
- Other loads: Other loads which may be required to be considered in design include impact loads (to consider collision from inside/outside the building on façade pan-

els), thermal loads (to cater daily and seasonal variations in ambient temperatures; this is particularly relevant for large façade systems), and loads due to forms of precipitation other than rain (such as snow or hail).

While the aforementioned actions are to be directly considered in structural design of façade systems, indirect actions that may cause loss of functionality during the service life of façade systems should also be considered. These include fatigue (primarily originating from wind loads) and creep (not very dominant due to relatively low static stress levels). Effects of fire are usually not considered as part of the main design of façade systems (from a strength perspective), and focus is kept on providing insulation from direct flame exposure to the extent possible. However, there seems to be a significant gap in considering structural behavior of façade systems (panels and cladding frame) at high temperatures.

Once the choice of panels has been made (SGU, DGU, ACP, etc.), the required thickness of the panels is calculated depending on the expected load (usually wind pressure governs). For example, ASTM E1300 (2016) provides guidance on calculation of thickness for glass panels of different kinds (SGU, DGU, etc.) along with minimum requirements and deflection calculations. NBC (2016) Section 6 also provides such guidance on selection of glass, its thickness, and allowable deflections. These usually depend on the thickness and aspect ratio of the glass. A similar guidance for ACP is not codified, but one can resort to the use of manufacturer datasheets which provide information regarding strength. Wood-based panels such as MDF can be dealt with using wood properties and calculation methods. Behavior of panels also depends on the support conditions—the manner in which they are connected with the cladding frame.

Cladding frame can be designed as a regular grid system comprising transoms and mullions. Design loads are calculated based on tributary areas from façade panels, as shown in Fig. 2.2. These areas are marked using 45° lines as shown; loads to be carried by respective transom/mullion can be calculated by multiplying the wind pressure acting on the panels with the tributary areas. As discussed earlier, self-weight and weight of façade panels also need to be considered. Predominantly, mullions are subjected to compressive forces and bending moments, while transoms are subjected to biaxial moments. Depending on the material, appropriate design methodology can be employed; for instance, if aluminum sections are being used, BS 8118 (1991) or Eurocode 9 Part 1-1 (2005) can be used. Design codes when using steel are also available. Some codes (e.g., NBC 2016) mention using the steel design code even when aluminum is being used (with appropriate material properties, of course). For Chlorinated Polyvinyl Chloride (CPVC) cladding frames, one may have to rely on manufacturer data and design calculations as codified approaches are limited.

Cladding frame is typically connected to the building through L-shaped metal brackets with one end connected to the cladding frame and the other connected to a building component (slab, beam, or column). The thickness, width, and diameter of fasteners are determined by the axial force, bending moment, and shear force acting on the bracket due to the actions on the façade (panels and cladding frame).

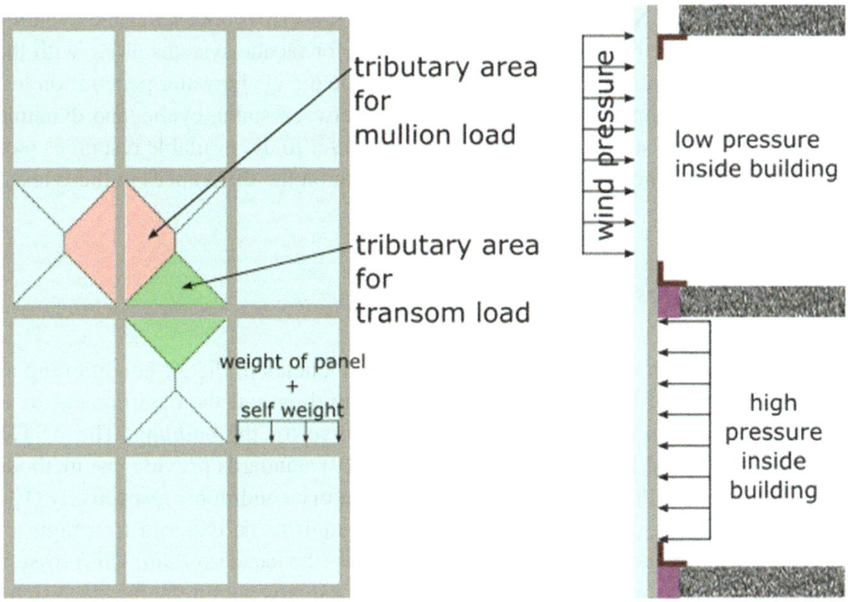

Fig. 2.2 Loads acting on façade panels and cladding frame. Tributary areas as shown are used to calculate loads acting on transoms and mullions. Wind pressure can act in either direction depending on the pressure conditions

2.3.2 Water Ingress

Water ingress through a façade system is highly undesirable as it can adversely affect living conditions inside the building. Even mild but prolonged dampness can lead to the growth of mold and render the living conditions unhygienic. Materials used in façade panels (glass, ACP, etc.) are typically waterproof by themselves. The manner in which these panels are connected with the main façade frame usually plays a deciding role in the water tightness of the overall system. Similarly, the locations where the façade assembly is connected with the building are also critical. The existing water penetration tests are of two types: static and dynamic. In static water penetration test, water is sprayed at the façade assembly at a certain flow rate at a uniform or cyclic static air pressure. The air pressure requirement can vary from building to building and even from location to location within the same building and should be chosen carefully. The ASTM E1105-15 (2015) gives a detailed set of procedures and performance criteria to be used for assessing the water ingress in façade systems under both uniform and cyclic static air pressure conditions. The specific ASTM standard for uniform testing is ASTM E331-00 (2016), while that for cyclic testing is ASTM E547-00 (2016). The ASTM E2268-04 (2016) provides methods for testing water penetration under dynamic air pressure conditions.

The Indian code, NBC (2016) Part 6, Section 8, prescribes both static and dynamic pressure water penetration tests to be carried out for façade systems along with the pass/fail requirements in different scenarios. While the cyclic water penetration test is mentioned to be optional, a clear distinction between static, cyclic, and dynamic water penetration tests is not available. One can refer to the available resources (see Lopez 2009) for a more comprehensive discussion on the differences in these tests.

2.3.3 *Air Ingress*

Air leakage through the façade can lead to inefficiencies in the air conditioning as it affects the rate of heat transfer between the building and the environment. It is also important from a weather-isolation perspective for the building. The ASTM E783-02 and ASTM E283/E283M-19 (2018, 2019) standards provide test methods for assessing air infiltration under field and laboratory conditions, respectively. The test usually requires the inside of the façade system to be exposed to a high air pressure, which creates a pressure differential across the façade system. This pressure differential causes the air to flow through the cracks/gaps within the façade system. The flow (discharge) of air through the façade is measured, and the system passes/fails the test depending on the acceptable upper limit for the flow rate.

The NBC (2016) Part 6 specifies the acceptable air leakage criteria for façade systems subjected to different pressure differentials and configurations. This code further specifies that a prior design wind load should be applied for a short duration before the actual test. It is to be noted that the wind pressure is the main structural load to be carried by most façade systems and that the functional requirements discussed here should be fulfilled under the design conditions. Both positive and negative pressures should be considered (from inside the façade system).

2.3.4 *Thermal Insulation*

Thermal insulation capability of a façade system affects the energy loss (or gain) by a building, which has a direct implication on its sustainability rating. There are different strategies for providing thermal insulation, but ventilated façade systems are quite common; they employ insulation materials in multiple layers (usually two) with an air gap in between. Within the air gap, hot air rises up and cold air enters from below acting as a naturally ventilated cooling system. The thermal insulation capability depends on three basic properties of a material, namely its density, specific heat capacity, and thermal conductivity. The general transient heat conduction equation is given by

$$\rho C \frac{\partial T}{\partial t} = \nabla \cdot (\lambda \nabla T), \tag{2.1}$$

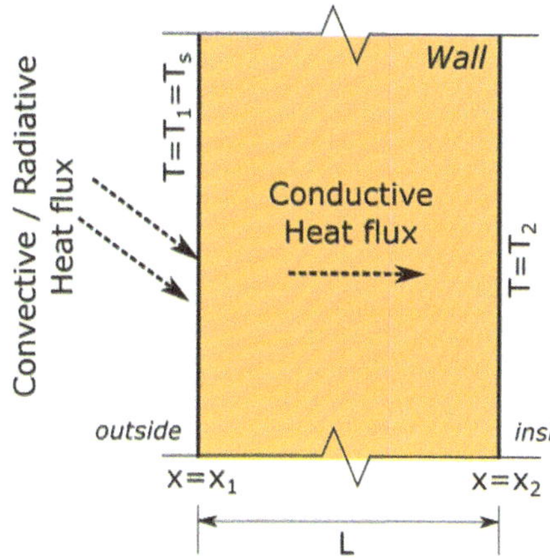

Fig. 2.3 Heat conduction through a wall

where ρ, c, and λ are the density (kg/m^3), specific heat (J/kgK), and thermal conductivity (W/mK), respectively, and T and t denote temperature and time, respectively. This equation hinges on energy balance and Fourier's law.

Consider a simple insulation wall/curtain wall as shown in Fig. 2.3. Such a situation can oftentimes be analyzed as a one-dimensional heat conduction problem given by

$$\rho C\frac{\partial T}{\partial t} = \frac{\partial}{\partial x}\left(\lambda\frac{\partial T}{\partial x}\right), \quad (2.2)$$

where the x direction indicates the direction along the thickness of the wall, L, as shown in Fig. 2.3. The effects of convective heat transfer (through the air movement) and radiative heat transfer (e.g., solar gain, discussed in Sect. 2.3.5) can be considered through a mixed type boundary condition given by

$$\dot{q}_{\mathrm{s}}'' = -\lambda\left.\frac{\partial T}{\partial x}\right|_{x=x_1} = \dot{q}_{\mathrm{c}}'' + \dot{q}_{\mathrm{r}}''. \quad (2.3)$$

Here, $\dot{q}_{\mathrm{c}}''$ and $\dot{q}_{\mathrm{r}}''$ are the convective and radiative heat flux defined, respectively, as

$$\dot{q}_{\mathrm{c}}'' = h_c(T_{\mathrm{g}} - T_{\mathrm{s}}), \quad (2.4)$$

and

$$\dot{q}_{\mathrm{r}}'' = h_r(T_{\infty}^4 - T_{\mathrm{s}}^4), \quad (2.5)$$

where h_c is the convective heat transfer coefficient, h_r is the radiative heat transfer coefficient (this involves the Stefan–Boltzmann constant, emissivity of the materials, and configuration factor), $T_{\mathrm{s}} = T_1$ is the temperature of the surface, while T_{g} and T_{∞} are the temperature of the gas and the radiation source; in many cases, T_{g} and T_{∞}

may be assumed to be the same. $\dot{q}_s''$ is the surface heat flux (going in or out of the material) in kW/m^2 and denotes the heat transfer rate per unit area perpendicular to the surface. The conductive heat flux, $\dot{q}_{cond}''$, relates to temperature through Fourier's law as

$$\dot{q}_{cond}'' = -\lambda \frac{\partial T}{\partial x}, \tag{2.6}$$

Fourier's law allows a simple correlation between the conductive heat flux and the temperatures, especially under steady-state conditions and constant thermal properties (when the temperature variation within the body becomes linear). For a linear temperature distribution,

$$\dot{q}_{cond}'' = -\lambda \frac{\partial T}{\partial x} = -\lambda \frac{T_2 - T_1}{x_2 - x_1} = \lambda \frac{\Delta T}{L}, \tag{2.7}$$

where $\Delta T = T_1 - T_2$ is the temperature difference between the inside and the outside and $L = x_2 - x_1$ is the thickness of the wall. From (2.7), the thermal conductivity, λ, can be interpreted as

$$\lambda = \dot{q}_{cond}'' \frac{L}{\Delta T}, \tag{2.8}$$

which leads to the definition of thermal conductivity as the rate of heat flow per unit area through a material of unit length ($L = 1$ unit) having a unit thermal gradient ($\Delta T = 1$ unit) with units as $(W/m^2)/(K/m)$.

For steady-state analyses, it is also convenient to define thermal resistance, R, as

$$R = \frac{L}{\lambda}, \tag{2.9}$$

which enables consideration of different layers of the insulation materials (e.g., a system of a wall and additional insulation material) in a way analogous to the consideration of electrical resistance following similar combination rules. Under certain simplifications, the convective and radiative boundary conditions can be defined in terms of corresponding thermal resistances and the heat transfer analysis can be simplified. For more details, see Wickström (2016).

In practice, the insulation capacity of façade panels is described in terms of three different but related physical/derived properties of materials:

- K-value: This is the thermal conductivity (W/mK), defined in (2.8). Materials with lower K-value are better insulators.
- R-value: This is the thermal resistance (m^2K/W), defined in (2.9). Materials with greater R-value are better insulators.
- U-value: This is the reciprocal of the thermal resistance, R, along with the effects of heat loss due to convection and radiation with units W/m^2K. Materials with lower U-value are better insulators.

U-value is most commonly used in assessing the insulation capabilities of façade materials primarily as it considers the effects of conduction, convection, and radiation, whereas R-value and K-value consider the effects of conduction only. These values also have relevance from a fire safety perspective as the main physical process in the response of façade systems to fire involves heat transfer.

The NBC (2016) Part 6 specifies the requirements on thermal insulation in terms of the U-value. Except for buildings in moderate climate, the maximum admissible U-value is given as 3.30 W/m^{2}K. For buildings in moderate climate, a greater U-value of 6.90 W/m^{2}K is allowed as the overall insulation requirements do not vary greatly due to change in seasons. Of course, the thermal insulation capability will also depend on the response of the façade materials to sunlight, which is discussed in Sect. 2.3.5. It can be noted that ultra-low U-values are possible by the use of multi-pane glazed units, discussed earlier. In fact, U-values in the range of 0.2–0.3 have been shown to make a building 'zero heating building' (Vanhoutteghem 2015; Kralj 2019).

An interesting point to note is that most materials that are good thermal insulators (low U-values) are poor in their fire performance as a majority of such products are petroleum-based polymers or foams (due to which they have high heat of combustion, low ignition temperatures, and low critical heat flux). Insulating materials can be classified into two categories – non-combustible and combustible. Non-combustible insulation materials include mineral wool (glass wool, stone wool, and ceramic wool). Combustible insulation materials are usually plastic foam materials such as phenolic foam, Expanded Polystyrene (EPS), Extruded Polystyrene (XPS), Rigid Polyurethane (PUR), Expanded Nitrile Rubber (ENR) and PIR (MacDonald, 2004).

Mineral wools are advantageous from a fire performance perspective due to their non-combustible nature. Glass wool and stone wool can be used up to a temperature of 250 and 850 °C, respectively, and are utilized in several fire protection applications. Ceramic wool can withstand up to 1200 °C but is quite expensive and is hence used in specific applications (e.g., internal lining of furnaces). Since mineral wools are open cell in structure, their porosity to gases is high, which necessitates the specification of packing densities (reducing porosity through compression) and/or use of foil facing and other mechanisms to provide gas barriers. The heat storage capacity (ρC) of mineral wools is generally lower than that of plastic foams, making them less effective for ambient thermal insulation applications. Additionally, handling of mineral wools requires care as they can cause irritation and itching.

Although the K-values of mineral wools and plastic foams mentioned earlier are in the same range (0.02–0.04 W/mK), the greater heat capacity of plastic foams makes them the materials of choice in providing ambient thermal insulation—one of the key requirements of a building façade. Unfortunately, all of these plastic foams melt in the range of 100–300 °C and can ignite when the temperatures reach around 400 °C making them supportive of fire growth. Low melting temperatures can cause early melting and dripping of these foams from their place of installation which can further invigorate a building fire. Thus, choice of materials for thermal insulation poses a challenge in optimizing between thermal comfort and fire safety.

2.3.5 *Response to Sunlight*

The earth receives about 1 kW/m^2 solar radiation, and if the area of the façade is large, it can have a significant effect on the overall ability of the façade to provide isolation from the environment. Additionally, large windows (transparent façade panels) may cause glare inside the buildings while shiny reflecting surfaces may cause glare outside the building. Façade panels are typically coated with solar control coatings to modulate the transmission and reflection of the incident sunlight as per the design requirements. There have been recent advances in exploration of the potential use of the façade surface for energy generation through solar photovoltaic cells (Unguresan 2017; Brito 2017). The effects of solar heat gain can be incorporated in the thermal analysis by considering $\dot{q}_r'' = 1\ kW/m^2$ as the radiative heat flux in the boundary condition mentioned in (2.3). The effects of fire on façade systems having solar photovoltaic cells are not yet known.

2.3.6 *Acoustic Insulation*

While the usual massive building materials (e.g., concrete and masonry) provide natural acoustic insulation, façade systems are usually built using thinner and low-mass materials. Thus, the acoustic insulation of the system should be checked and ensured. The general requirements of sound insulation are described by NBC (2016) Part 8. ASTM E966-18a (2018) also provides guidance on measurement of sound attenuation of building façades and façade elements.

2.4 Main Parameters Considered for Design

The overall design process of a façade system can be broken into four components:

- Aesthetics: Consideration of aesthetics is perhaps the precursor to any façade design as it intends to provide a face and character to a building.
- Design for thermal comfort: This entails considering the climate type, choosing the desired U-value, and selecting appropriate façade panel materials. One may also consider a regular or a ventilated façade system at this stage.
- Structural design: At this step, the strength and stability of the façade panels are ensured along with the design of the cladding frame members and their connection details. The primary design load is usually wind, especially for non-load-bearing systems, as discussed earlier.

- Consideration of fire safety: This aspect is usually considered only at panel levels, wherein the fire ratings of the façade panels are taken into account. However, the overall fire behavior of the façade as a system can be quite different when compared to that of individual components, and hence, it is advisable to consider fire safety as a separate design step. A detailed discussion on component vs. assembly vs. system-level fire performance and testing is presented in Chap. 4.

2.5 Firestop

Firestop systems are designed to ensure compartmentation during the event of a fire and are an essential part of passive fire protection strategy of a building. Any gaps, either existing ones after construction (e.g., gap between edge of the slab and façade) or the ones which can form during a fire (e.g., due to melting of a plumbing pipe), are to be considered during the design of firestop systems. ASTM E283/283M-19 (2017) provides guidance on performance of through-penetration firestop systems, which is measured in terms of the following ratings:

- T-rating: This is a thermal rating, measured as the time in hours up to which the temperature of the non-fire side of the penetration remains below a defined threshold level (typically in the range of 140–170°C). Some codes refer to this as an insulation rating.
- F-rating: This is a flame rating, measured as the time in hours up to which the firestop can withstand fire before permitting passage of hot gases. Some codes refer to this as an integrity rating, and some require F-rated openings to withstand a hose stream test also.
- L-rating: This is a smoke rating, measured as the amount of air or cold smoke (in cubic feet per minute or cubic meter per minute) that can leak through a firestop with a certain pressure differential being maintained for the stipulated test duration.
- W-rating: This is a relatively new rating which quantifies the ability of a firestop to resist passage of water and originated after concerns raised by building owners related to water ingress problems.

Given that most deaths in fire accidents are attributed to smoke, the L-rating seems to be fairly important. However, the use of cold smoke in the standard tests renders it somewhat non-conservative. This is due to the fact that during a fire, there will be some amount of hot movement of the local area and there is a chance that before the F-rating of a system is achieved, it may allow passage of a non-trivial volume of smoke.

Chapter 3
Materials Used and Their Properties

This chapter presents an overview of material properties and choice of materials for different components of façades. Further, relevant combustion properties of the material used in façade systems are discussed. Materials used in the case studies considered in Chap. 5 are discussed in greater detail.

3.1 Cladding Frame

Aluminum and steel are perhaps the most used materials for the cladding frame. While aluminum is lightweight and is especially suited for non-structural façade systems, it exhibits relatively poor fire behavior compared to steel as it melts at around 660 °C, while steel loses about half of its strength around this temperature. While better in fire performance, steel has the disadvantage of being heavier. As can be seen from Table 3.1, the specific strength (strength-to-weight ratio) of steel is about half of that of aluminum. Aluminum also has greater thermal conductivity (about six times) compared to steel, which is a disadvantage not only from fire perspective but also from ambient thermal requirements discussed earlier.

Modern windows and some façade systems utilize PVC and its variants (Unplasticized Polyvinyl Chloride (UPVC) and CPVC) in the frame members. Initially, this family of materials replaced plumbing pipes but in recent years, their popularity in window frames has increased significantly due to their low thermal conductivity and ease of forming. This ease of forming allows creation of complex and intricate cross sections that can enable more efficient thermal design of windows and façade systems, which cannot be achieved through metals (steel or aluminum). The better thermal behavior of the PVC family comes at a cost—its specific strength is much inferior to that of aluminum even though the unit weight is about half of that of aluminum. Another disadvantage of PVC is its susceptibility to fire as it is a combustible

G. Srivastava and P. D. Gandhi, *Performance of Combustible Façade Systems Used in Green Building Technologies Under Fire*, Springer Transactions in Civil and Environmental Engineering, https://doi.org/10.1007/978-981-16-3112-2_3

Table 3.1 Properties of materials typically used in cladding frame

Property	Steel	Aluminum	PVC	CPVC
Density (kg/m^3)	7850	2700	1400	1520
Tensile strength (MPa)	410	290	51	53
Specific strength (m^2/s^2)	0.05	0.11	0.04	0.03
Thermal conductivity (W/mK)	43	237	0.147	0.137
Concern during fire	Mild	Moderate	High	High

material and requires characterization of flame spread, ignition temperature and other combustion properties for design although the 'self extinguishing' property of PVC and CPVC gives some relief. Further, design approach using steel and aluminum is fairly standardized (e.g. Eurocode 3 Part 1-1 (2005) and Eurocode 3 Part 1-2 (2005) provide material properties and guidance on design calculations for steel frames under general and fire conditions, respectively and Eurocode 9 Part 1-1 (2005) and Eurocode 9 Part 1-2 (2005) provide the same for aluminum frames). For PVC and CPVC, one has to rely on manufacturer data sheets though other sources are available (Tittow 1986). Tittow (1986) also provide a list of standards for specifications of different PVC properties.

3.2 Thermal Insulation

A large number of materials with low conductivity (K-value) find applications in exterior building insulation with the intent to reduce the overall energy loss/gain by the building. These include brick masonry, concrete, rockwool or slag wool, wood, stone, petroleum-based polymers (polystyrene, polyurethane), mineral boards, etc. Some good resources on the assessment of thermal performance of such materials are available (BNL 1976; Strother 1990; Nelson 1996).

The discussion here will be restricted to PIR, which has been used in two of the case studies (Cases 6 and 7) presented in Chap. 5. PIR is typically used as a rigid thermal insulation and is manufactured from methylene diphenyl diisocyanate and a polyester-derived polyol. Its thermal conductivity (K-value) is extremely low at 0.023 W/mK which makes it suitable for thermal insulation applications. Its fire behavior has been of concern though. Hidalgo (2017) studied the fire behavior of PIR to characterize a critical temperature-based performance-based design criteria for insulation boards and suggested a critical temperature of 300 °C for PIR.

PIR boards used in the current context were analyzed for thermal decomposition and combustion behaviors using Thermogravimetry Analysis (TGA) and Cone Calorimeter. Figure 3.1a presents the mass loss curves of PIR at different heating rates. It can be observed that substantial mass decomposition takes place around 300 °C and PIR leaves a residue of about 20% after decomposition. Heat release

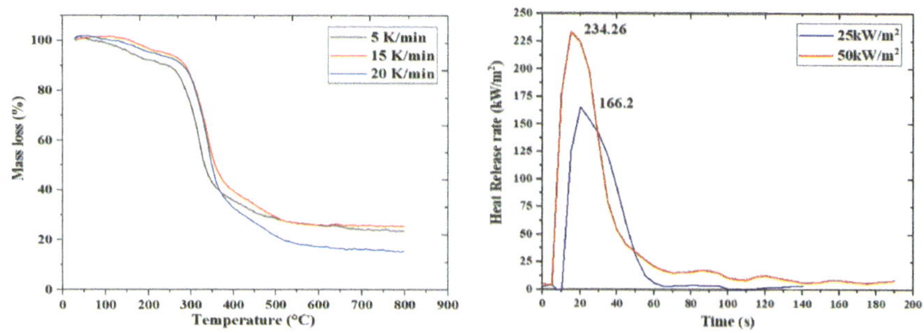

(a) Mass loss behavior.

(b) Heat release rate.

Fig. 3.1 Mass loss and heat release rate of PIR

(a) Before experiment.

(b) After experiment.

Fig. 3.2 Images of PIR samples before and after cone calorimeter experiment at 25 kW/m^2 exposure. Charred burning behavior of PIR can be seen. The unburnt area at the periphery was due to the shielding effect of the cone calorimeter sample holder

rate, as characterized by cone calorimeter, is shown in Figure 3.1b for two different incident heat flux levels of 25 and 50 kW/m^2. Figure 3.2 shows images of the PIR samples before and after a cone calorimeter experiment. A color gradation from yellow (unburnt) to orange-black (partially burnt) and black (charred burnt) can be observed.

A summary of the combustion properties quantified through cone calorimeter experiments is given in Table 3.2. It can be observed that PIR ignites readily (within 2 s even at a moderate flux exposure of 25 kW/m^2) and demonstrated a large peak Heat Release Rate (HRR) shortly after ignition and leaves a black char when burning. During thermal degradation, discoloration of PIR samples was observed wherein their color changed from yellow to orange-brown to black. As will be discussed in Sect. 5.8, PIR can play a significant role in fire growth. It was also observed that PIR did not burn significantly at incident flux level of 10 kW/m^2 indicating a critical heat flux in the range of 10–25 kW/m^2.

Table 3.2 Combustion properties of PIR

Incident flux (kW/m^2)	Time to ignition (s)	Peak HRR (kW/m^2)	Peak CO yield (kg/kg)	Total heat release (MJ/m^2)
25	8	166	1.15	5.2
50	2	234	4.82	7.9

3.3 Façade Panels

Multiple varieties of materials are used in façade panels. This section discusses some of the relevant ones.

3.3.1 *ACP*

ACP is a member of the more generic metal composite panels in which a 'core' material is sandwiched between two metal sheets. The most common core material used in ACPs is Polyethylene (PE) or Polyurethane (PU) . In many cases, recycled PE is used as it serves to partially alleviate the generic low density plastic waste problem. PE is usually highly combustible; the metal sheets provide some hindrance to direct heat exposure of the core material but that only delays the participation of ACPs in fire events. Recently, Fire Resistant (FR) varieties of ACPs have become common which incorporate an inorganic mineral infill (e.g. magnesium hydroxide) causing substantial improvement in their fire performance.

Being a layered composite, it is customary to determine few combustion properties (such as calorific value) by testing individual components (top paint layer, primer, adhesive film, and core material) and then combining the results for ACP. Certain properties, such as time to ignition and critical heat flux, depend on the thickness of the metal sheets and the overall ACP and hence, must be determined for the full system. The University of Queensland (University, 2021) has developed an extensive library of combustion properties of different types of cladding materials (primarily ACPs) which is an excellent resource in this subject.

Figure 3.3a shows the mass loss behavior of the core PE of an ACP. It can be observed that PE loses most of its mass between 400 and 600 °C and leaves about 20% residue after complete decomposition. A somewhat higher residue is due to the fact that ACPs used herein were made of recycled PE and hence contained some impurities. Virgin PE is expected to leave less than 10% residue.

Figure 3.3b shows heat release rate of the PE core and ACP sheets of 3 mm thickness subjected to a 50 kW/m^2 incident flux in cone calorimeter. It can be observed that aluminum sheets (0.25 mm thick) provide an initial inhibition to the onset of fire in the ACP and release of heat is delayed subsequently. A small peak observed in case of ACP before the major one corresponds to the combustion of Polyvinylidene

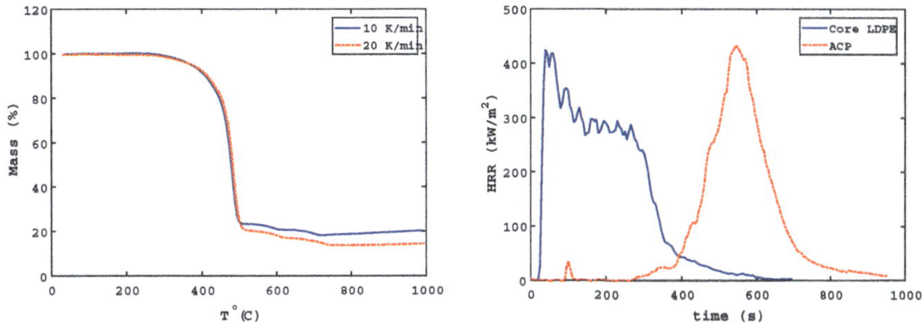

(a) Mass loss behavior. (b) Heat release rate.

Fig. 3.3 Mass loss and heat release rate of PE and ACP

(a) Sample being removed. (b) View of the polymer core.

Fig. 3.4 Images of ACP samples after cone calorimeter experiment at 25 kW/m^2 exposure. Leftover scraps of the PVDF coating on the top aluminum sheet can be seen. Neither of the aluminum sheets burnt. The polymer core melted and foamed before burning. Black residue near the edges formed due to incomplete combustion and shielding effect of the sample holder

Table 3.3 Combustion properties of ACP at 50 kW/m^2 exposure

Thickness (mm)	Time to ignition (s)	Peak HRR (kW/m^2)	Peak CO yield (kg/kg)	Total heat release (MJ/m^2)
3	266	434.1	1.07	79.2
4	414	588.6	2.34	103.5

Difluoride (PVDF) coating on top of the ACP sheet (typically done to provide color and weather protection to the surface finish). Figure 3.4 shows images of the ACP samples tested in the cone calorimeter.

A summary of the combustion properties quantified through cone calorimeter experiments is given in Table 3.3.

3.3.2 Fiberboards (MDF)

Fiberboards are available in different varieties—e.g., mineral fiberboards, glass fiberboards, wood fiberboards, and guidance on their use and material properties is available (Reid 2000). In the current context, wooden medium density fiberboards (MDF) were used. These comprised of 92% pinewood, 7% phenol formaldehyde resin, and 1% paraffin wax. The overall density of MDFs was about 700 kg/m^3. Being wood-based, the behavior of MDFs was similar to that of pinewood that demonstrates charring behavior as well as a double peak heat release rate behavior due to decomposition of cellulose and hemi-cellulose at different times during the combustion process (Nakrani 2020).

3.3.3 Glass

Glass is a non-combustible constituent of any façade system and is available in a variety of forms, shades, and grades—which are chosen based on the application. As discussed in Chap. 2, multi-pane glazing units are also used. In the experiments presented in Chap. 5, SGUs were used. Glass panes were typically of 6 mm thickness and were of float and toughened type. Details on different properties and design stipulations for glass can be found in the literature (Patterson 2011).

3.4 Firestop

Firestop systems are used to ensure proper compartmentation across enclosures of a building. A building typically has many services that cross compartment boundaries horizontally or vertically. These include Heat Ventilation and Air Conditioning (HVAC) ducts, plumbing pipes, electrical conduits, and service shafts; some of these are shown in Fig. 3.5.

In the context of façades, firestop systems are required at edge of the slab gaps between the floor slab and the façade, typically called the safing area. Figure 3.6 shows a firestop installed in the safing area. Firestop systems typically comprise of a fire insulation—usually mineral wool of some kind installed at varying levels of packing density. The packing density typically governs the porosity of the firestop; greater porosity (lower packing density) provides lesser barrier to hot gases while lower porosity (greater packing density) provides better barrier to hot gases. In the experiments presented in Chap. 5, mineral wool packing densities in the range of 96–128 kg/m^3 were used. Greater densities also allow better performance of the firestop system in case of hot movements of the façade frame during a fire. The top of the firestop installation is sometimes covered with metal plates (e.g. stainless steel) or propriety smoke and fire seals (wet systems). Foil-faced systems that are

(a) Electrical conduit firestop.

(b) Plumbing pipe firestop.

Fig. 3.5 Firestop systems typically used for passage of electrical conduits and plumbing pipes. During the event of a fire, these systems activate thermally and expand to fill the gap to prevent passage of hot gases

Fig. 3.6 Safing firestop

pre-compressed to the desired packing density and can be installed directly on site are also available.

Safing area firestop systems are not complete without suitable treatment of the spandrel area from the inside. This is also achieved through similar materials—mineral wool and its forms–with appropriate packing density. When greater packing density is to be used, use of stainless steel backpan in the spandrel area is advisable.

3.5 Connections and Other Materials

Connections between different constituents of a system require special attention as they can easily become the weakest links of the system. This section discusses some necessary connections within façade systems.

3.5.1 Frame-Building Connection

Cladding frame (mullions and transoms) is usually connected at regular intervals to the building columns and slabs at all floor levels through metal brackets. These brackets are usually L-shaped where one leg is connected to the building through anchor fastener and other leg is connected to the cladding frame. Commonly used metals are steel (mild steel as well as stainless steel) and aluminum. Thickness and width of the metal brackets are decided based on the structural load calculations (loads transferred by the cladding frame to the building).

3.5.2 Frame-Panel Connection

Façade panels (glass, ACP, etc.) are connected to the cladding frame through either pressure plates or mechanical fasteners. The quality and type of this connection governs performance parameters such as air ingress, water ingress, etc., of the façade system. In stick systems, use of pressure plates is prevalent. In semi-unitized systems, glazing units are fabricated at factory and are connected to the cladding frame on-site using either of the two methods, use of mechanical fasteners being more common. Unitized systems are completely built in factory and hence give the best performance for façade requirements; unitized units are connected directly to the building through mechanical fasteners.

3.5.3 Panel Adhesives

There are two locations where adhesives are typically used in façade systems. First is the façade panel—cladding frame connection, if done through pressure plates and silicon sealant. Second is the adhesive used inside metal composite panels (e.g. ACP) to sandwich metal sheets to the inner polymer core.

Chapter 4
Rationale of Existing Test Methods

In the USA, concerns about energy efficiency led to introduction of insulation materials in building facades. Many of these materials were flammable. This raised a concern among the fire safety experts in the USA regarding the fires that may involve facades. In this context, the Society of Plastics Industries (SPI), now Plastics Industry Association, sponsored development of a test to determine the fire performance of a composite façade system consisting of combustible thermal insulation. A photograph of the test developed is shown in Fig. 4.1.

The test fixture consisted of an outside two-story building that was 24 ft (7.3 m) high, with floor heights of 12 ft (3.7 m). The fire exposure represented a room and content fire that break out of a window opening with the flames extending to the façade exterior. The physical scale of testing required the test to be conducted outdoors. The fire performance metrics included the following:

- Vertical and lateral flame propagation over the exterior face of the wall assembly;
- Vertical flame propagation within the combustible core, air cavities, or within combustible components from one story to the next;
- Vertical flame propagation over the interior surface of the wall assembly from one story to the next; and
- Lateral flame propagation from the compartment of fire origin to adjacent compartments or spaces.

The research demonstrated that the test was able to develop data on the aforementioned fire performance metrics for façades with combustible components. The test was adopted for regulating façade systems in 1988 in the USA by the Uniform Building Code (UBC).

Since the development of this test, other weaknesses in high-rise buildings became apparent from a collection of high-profile fires. One such fire occurred at the First Interstate Bank (Los Angeles) on May 4, 1988. The building was 62 floors with concrete and steel construction and a glass façade. The fire spread from the floor

G. Srivastava and P. D. Gandhi, *Performance of Combustible Façade Systems Used in Green Building Technologies Under Fire*, Springer Transactions in Civil and Environmental Engineering, https://doi.org/10.1007/978-981-16-3112-2_4

Fig. 4.1 Photograph of SPI façade fire test

of fire origin (12th floor) up to the 16th floor. The transmission of flames from the floor of fire origin to upper floors occurred after the 12th floor window broke and the flames progressed to the façade and the windows above (Routley 1988; Nelson 1988). There were several possible means of fire progressions to the upper floors based upon post-fire evidence (e.g., photographs, video tapes, fire department observations):

- Extension of flame from the floor of fire origin to the window area on the floor above;
- Failure of the suspended ceiling and glass spandrel panel, followed by extension of flames around the end of the floor slab;
- Flame transmission between the glass spandrel panel and the end of the floor slab. In this instance, the actual transmission of fire may have resulted from failure of the gypsum board panel extending from the windowsill on the floor above the fire floor level, failure of the aluminum windowsill or mullions, or failure of the glass spandrel panel; and

- Passage of flame through spaces around pipes and other penetrations of the floor slab. This route for the flame is evidenced by burn traces in the core area.

The first item here relates to the contribution to fire growth by the façade system. The second and third items relate to lack of fire protection in the gap between the floor slab and the glass facade (safing area), and the fourth item relates to a lack of adequate fire stopping around the through-penetrations created by utilities such as HVAC ducts and plumbing.

The fire protection gaps in the aforementioned items have also been identified in other recent fires such as the Grenfell Tower fire in the UK (Grenfell 2019) and the Lotus Building fire in India.

4.1 Fire Exposure to Building Façades

Figure 4.2 illustrates (Development of a European Approach 2018) a multi-story building where a fire originates on the Second Floor (SF). A fire scenario may develop with a fire starting in a room. If during the fire, a window breaks and provides oxygen to support the combustion, flashover conditions will occur and flames will be ejected from the window. At this point, the fire size inside the room of fire origin will be constrained by the oxygen available through the window opening. The buoyancy-driven fire plumes on the exterior of the façade heats and ignites the cladding leading to further progression of the fire. Important concepts are presented in the following sections to facilitate better understanding of the fire phenomenon, the research conducted, and standardized testing used for regulating façade systems relative to their fire performance.

4.1.1 Heat Release

Once the fire develops, the temperature in the room of fire origin increases as the fire grows. The breaking of a window from the elevated temperature (Pagni 1991) occurs when the room temperature reaches approximately 450 °C. The additional infusion of air into the room of fire origin may bring the room to flashover with flames and fire plume emerging from the window opening. The minimum HRR to cause flashover, Q_{fl}, is calculated using semiempirical correlations based upon achieving a temperature of 500–600 °C in the enclosure hot gas layer. Three correlations attributed to Thomas (1981)—(4.1), Hurley (2016)—(4.2), and Yanagisaw (2007)—(4.3) are, respectively:

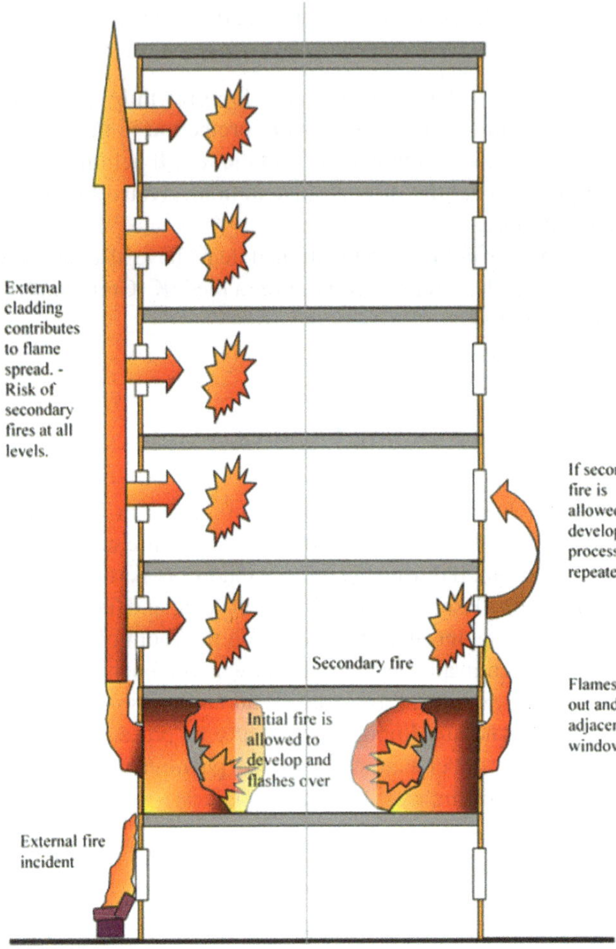

Fig. 4.2 Schematic of multi-story fire involving building façade (Development of a European Approach 2018)

$$Q_{\mathrm{fl}} = 7.8A_{\mathrm{T}} + 378A_{\mathrm{v}}\sqrt{H_{\mathrm{v}}}, \tag{4.1}$$

$$Q_{\mathrm{fl}} = 750A_{\mathrm{v}}\sqrt{H_{\mathrm{v}}}, \tag{4.2}$$

$$Q_{\mathrm{fl}} = 150A_{\mathrm{v}}\sqrt{H_{\mathrm{v}}}\left(\frac{A_{\mathrm{T}}}{A_{\mathrm{v}}\sqrt{H_{\mathrm{v}}}}\right)^{0.4}. \tag{4.3}$$

Here, Q_{fl} is the HRR required for flashover (kW), A_{v} is the window opening area (m^2), H_{v} is the window height (m), and A_{T} is the total wall and ceiling area (m^2). A ventilation or opening factor, F_{v}, used in many calculations pertaining to HRR, is defined as

$$F_{\mathrm{v}} = \frac{A_{\mathrm{v}}\sqrt{H_{\mathrm{v}}}}{A_{\mathrm{T}}}. \tag{4.4}$$

Sometimes, a definition of opening factor unscaled with the total area, A_T, is used as $F_v = A_v\sqrt{H_v}$.

The equations are semiempirical and are derived from analysis of enclosure fire data. (4.1) was developed from real-scale fire experiments. (4.2) is a simplified equation using only the vent dimensions and the assumption that $F_v^{-1}A_T/A_v\sqrt{H_v} \approx 50$. (4.3) is based on small-scale experiments with 0.5×0.5 ×0.5 m fire enclosures. The smaller, scaled experiments have recently been used to characterize fire exposure to façades with fires ejected from a window opening.

A more complete discussion on estimation of HRR for occurrence of flashover in an enclosure is provided by Walton (2016). If the heat input is greater than Q_{fl}, there will be some unburned fuel exiting from the window opening. The excess fuel will result in flames external to the room of fire origin. The excess HRR in the fire plume ejected from the window is given by

$$Q_{ex} = Q_{in} - Q_{fl}. \tag{4.5}$$

The heat input, Q_{in}, for tests performed with gaseous fuel may readily be calculated based upon flow rate and the fuel's heat of combustion. For solid fuels (e.g., wood cribs and others), Q_{in} may be measured using a HRR calorimeter under well-ventilated conditions.

It needs to be emphasized that Q_{ex} is the HRR from combustion of excess fuel outside the room of fire origin and is not the total HRR of the fire plume ejected from the window opening. The total HRR, Q_t, is the sum of the convective HRR of the hot gases and the HRR from combustion of excess fuel in the fire plume as

$$Q_t = Q_c + Q_{ex}. \tag{4.6}$$

The convective HRR of the fire plume hot gases, Q_c, can be calculated as

$$Q_c = \dot{m}_g C_g \left(T_g - T_\infty\right), \tag{4.7}$$

where $\dot{m}_g$ is the mass flow rate of hot gases out of the window, C_g is the specific heat of the gases, and T_g is the gas temperature. In experiments, Q_c may be calculated by measuring temperature and velocity at the window opening. There are several engineering correlations available for pre-flashover and post-flashover estimation of temperature exiting the window (Walton 2016).

For a fully developed under-ventilated condition in the room of fire origin, one may consider that the incoming air is completely consumed by combustion. Under this condition, the mass of air inflow rate is:

$$\dot{m}_{in} = 0.5A_v\sqrt{H_v}. \tag{4.8}$$

The HRR from the consumption of oxygen from the incoming air may be calculated as

$$Q_{st} = 0.5Y_{ox}\Delta H_{ox}A_v\sqrt{H_v}, \tag{4.9}$$

where Y_{ox} is the mass concentration of oxygen in air and ΔH_{ox} is the heat released per unit mass of oxygen consumed and approximated as 13.1 MJ/kg for a wide range of fuels (Huggett 1980). Thus, the HRR from complete combustion of incoming air may be estimated as

$$Q_{\text{st}} = 1500 A_{\text{v}} \sqrt{H_{\text{v}}}. \tag{4.10}$$

Q_{st} represents the maximum HRR that can be generated due to the constraints of available oxygen. It may be noted that (4.2) is $0.5 \times Q_{\text{st}}$ in (4.10).

Many scaled experiments in façade research are performed with $Q_{\text{in}} > Q_{\text{st}}$. This provides the advantage of near constant temperature in the fire compartment since combustion energy output is maximized within the compartment and any excess fuel is combusted outside of the fire compartment. This scenario provides consistent test conditions to study fire exposures from flames ejected from a window opening with well-established neutral pressure plane. In these scaled experiments, excess HRR has been calculated as

$$Q_{\text{ex}} = Q_{\text{in}} - Q_{\text{st}}. \tag{4.11}$$

A non-dimensional HRR, given in (4.12), based upon Froude number (Zukoski 1981; Heskestad 1986) has been found to be very useful for studying fire plumes and will be used for non-dimensional correlations for temperature and flame heights.

$$\overset{*}{Q} = \frac{Q}{\rho_\infty C_{p,\infty} T_\infty \sqrt{g} D^{2.5}}, \tag{4.12}$$

where Q is the HRR, ρ_∞ is ambient air density, C_p is the specific heat, T_∞ is the ambient air temperature, g is acceleration due to gravity, and D is the characteristic dimension of the fire.

4.1.2 Virtual Origin

All temperature and flame height correlations for fire plumes are based upon Morton's theory of a virtual origin for the thermal plume source (Morton 1956). Thus, an estimation of the virtual origin is essential in using these correlations. Heskestad (1983) developed a model for the virtual origin of axisymmetric fire plumes with the approximation given by

$$\frac{z_0}{D} = -1.02 + 0.083 \frac{Q_{\text{c}}^{0.4}}{D}, \tag{4.13}$$

where z_0 is the virtual origin (m), Q_{c} is the convective HRR (kW), and D is the length scale (e.g., diameter) (m).

For façade fires, the fire plume outside the window is a half-axisymmetric thermal plume. Tang (2012) from their scaled experiments developed an estimation of the virtual origin given as

$$\frac{z_0}{l_1} = -4.14 + 2.2\frac{Q_c^{*\,0.4}}{l_1}, \tag{4.14}$$

where l_1 is the characteristic length scale. Using ambient temperature as 295 K in (4.12) may be reduced to the same form as (4.13) as

$$\frac{z_0}{l_1} = -4.14 + 0.133\frac{Q_c^{0.4}}{l_1}. \tag{4.15}$$

4.1.3 *Window Opening*

The HRR external to the window opening is used to develop generalized correlations for plume temperatures and heat flux to the façade exterior. Yokoi (1960) identified the ventilation aspect ratio defined as

$$n = \frac{W_v}{0.5H_v}. \tag{4.16}$$

Yokoi's work demonstrated how the ventilation aspect ratio influences the trajectory of plumes exiting a window opening. Narrower windows (lower value of n) lead to greater plume exit velocity and result in the plume trajectory initially being further away from the façade wall. Yokoi's results indicated that attachment of the flames was more likely for $n > 3.6$. Using the window aspect ratio, n, Yokoi was able to calculate the minimum spandrel height needed to prevent breakage of glass above the room of fire origin. In this analysis, Yokoi developed important correlations for fire plume temperatures using the vent dimensions to identify an equivalent window scale based upon the window area.

Experiments by Peng (2016) showed the influence of the ventilation aspect ratio, n, on the trajectory of the fire plume emerging from an opening. The window dimensions and the aspect ratio are presented in Table 4.1.

Table 4.1 Data from Peng (2016)

Test #	W_v (m)	H_v (m)	n
1	1.9	1.1	3.45
2	1.4	1.4	2.0
3	0.9	1.9	0.95

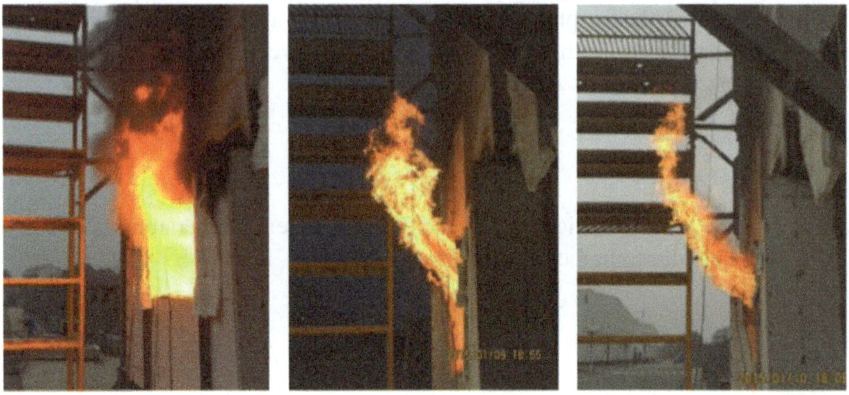

(a) Test 1, $n = 3.45$ (b) Test 2, $n = 2.0$ (c) Test 3, $n = 0.95$

Fig. 4.3 Turning of flames from the window (Peng 2016)

The photographs from this testing of the fire plume trajectory for these three different aspect ratios are presented in Fig. 4.3.

As n increases, the fire plume trajectory from the window becomes closer to the façade wall above it. The thermal exposure from the fire plume is also influenced by the window aspect ratio and is expected to be higher with an increase in n due to increased convective and radiative heat transfer from the fire plume to the façade.

Lee (2012) revisited Yokoi's analysis and developed a rationale for a corrected length scale l_1 for convective outflow of gases from a window opening as

$$l_1 = \left(\frac{\rho_g}{\rho_\infty} A_v \sqrt{H_v} \right)^{0.40}, \tag{4.17}$$

where ρ_g is the density of hot gases exiting the window and ρ_∞ is the ambient air density. Since density ratio (ρ_g/ρ_∞) is a weak function of temperature, l_1 can be simplified as a scale based on the window opening as

$$l_1 = \left(A_v \sqrt{H_v} \right)^{0.40}. \tag{4.18}$$

Lee (2012) defined another length scale, l_2, to represent the balance of momentum of gases exiting the window and upward buoyancy of the hot gases as

$$l_2 \propto \left[\frac{\rho_g}{\rho_\infty} \left(1 - \frac{z_0}{H_v} \right) \right] \left(A_v H_v^2 \right)^{0.25}, \tag{4.19}$$

where z_0 is the location of the neutral pressure plane. For fully under-ventilated fires (i.e., $Q_{in} > Q_{st}$), the exit temperature and the neutral pressure plane do not change with Q_{in}. Then,

$$l_2 \propto \left(A_v H_v^2\right)^{0.25}. \tag{4.20}$$

The length scale l_2 is similar to window aspect ratio, n, used by Yokoi (1960), as it influences the fire plume trajectory out of the window. Higher values of l_2 denote greater momentum of hot gas outflow, and therefore, flames from the window opening are anticipated to move further horizontally before turning upward due to buoyancy forces.

The scales l_1 and l_2 have physical implications as dimensions of a burner with sides l_1 parallel to the window and l_2 perpendicular to the façade. This theoretical burner would be located at the neutral pressure plane with a HRR of Q_{ex} (4.11). This is illustrated in Fig. 4.4.

A consequence of the new length scale, relevant for window outflow, is a modification of the non-dimensional HRR defined in (4.12) by replacing the length scale D with l_1, as

$$\overset{*}{Q} = \frac{Q}{\rho_\infty C_{p,\infty} T_\infty \sqrt{g} l_1^{2.5}}, \tag{4.21}$$

The length scales l_1 and l_2, and the ventilation ratio, n, are important in comparing the fire performance tests used for façades.

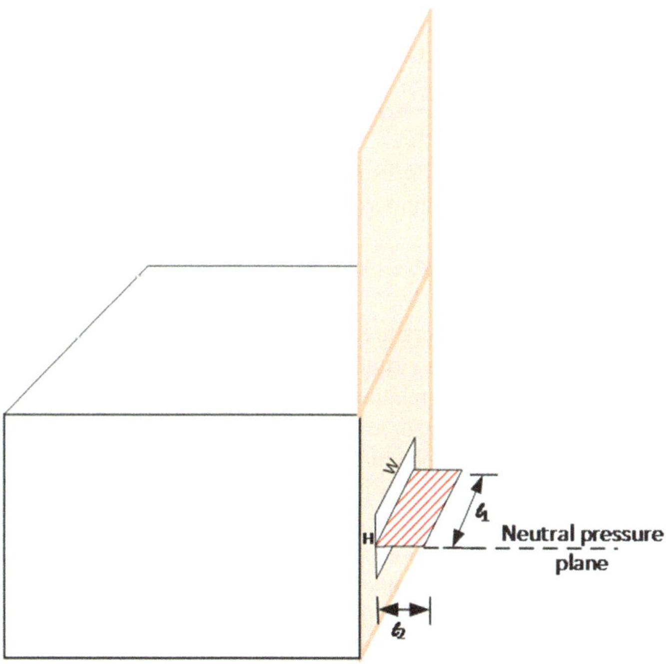

Fig. 4.4 Illustration of window opening length scales

4.1.4 Temperature on Façade

Yokoi (1960) conducted a number of experiments to study the fire plume ejected from a window opening and its potential for fire spread on building façades. The experiments resulted in key correlations of thermal plume temperatures and have been widely referenced in the literature. Yokoi used a length scale, r_0, based on the window opening for developing a non-dimensional correlation for temperature. The length scale r_0 was based on the equivalent diameter as

$$r_0 = \sqrt{\frac{W_v H_v}{2\pi}}. \tag{4.22}$$

The non-dimensional centerline temperature, θ, on the façade wall at a distance z, above the virtual origin, z_0 was then given as

$$\theta = \frac{\left(\frac{\Delta T_z}{T_\infty}\right) r_0^{5/3}}{\left(\frac{Q^2 T_\infty}{C_p^2 \rho_z^2 g}\right)^{1/3}} = f\left(\frac{z}{z_0}\right). \tag{4.23}$$

Using (4.12), the non-dimensional centerline temperature may be expressed as

$$\theta = \frac{\left(\frac{\Delta T_z}{T_\infty}\right) r_0^{5/3}}{\overset{*}{Q}^{2/3}} = f\left(\frac{z}{z_0}\right). \tag{4.24}$$

Beyler (1986) analyzed Yokoi's results and developed an engineering correlation for the centerline plume temperature as

$$\Delta T_z = 24.6 Q^{2/3} Z^{-5/3}, \tag{4.25}$$

where ΔT_z is the centerline temperature increase (°C) relative to ambient temperature and z is the height (m) above the virtual origin, z_0.

Figure 4.5 shows Yokoi's correlation excerpted from Lee (2012). Lee identified three distinct regions: Region 1 ($z/r_0 < 1$) is immediately above the window opening, where the non-dimensional temperature does not change significantly; in Region 2 ($z/r_0 > 1.5$), θ varies inversely with height; in Region 3 ($z/r_0 > 10$), θ varies as $(z/r_0)^{-5/3}$).

Lee revisited Yokoi's correlation and proposed two changes to Yokoi's non-dimensional temperature correlation. The first was to develop length scales that represent the flow from window openings. These were described as l_1 and l_2 by (4.18) and (4.19), respectively. The second modification was to replace local den-

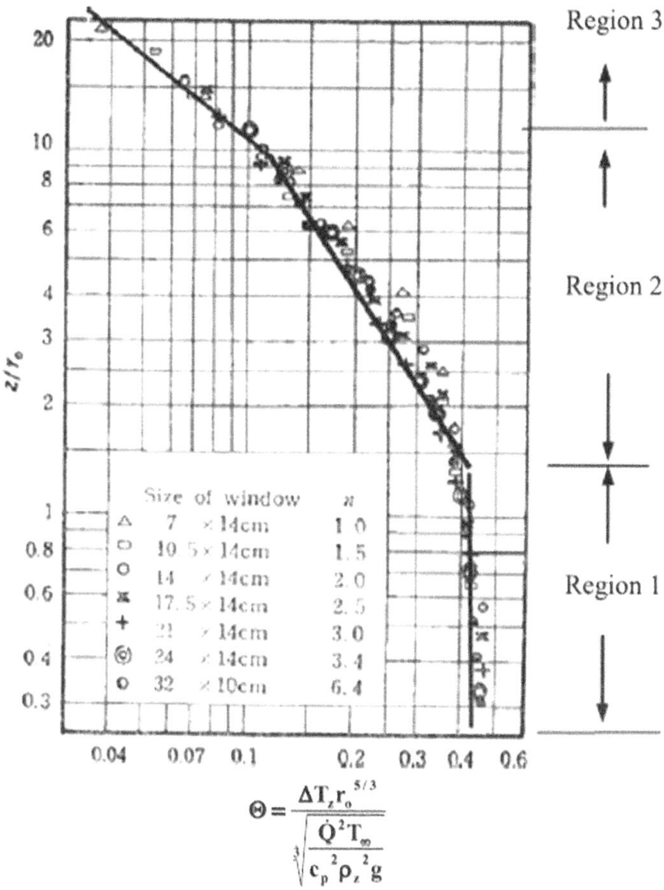

Fig. 4.5 Temperature correlation by Yokoi (1960)

sity ρ_z with ambient density ρ_∞. This results in modification of (4.23) and (4.24), respectively, as

$$\theta = \frac{(\Delta T_z)\, l_1^{5/3}}{\left(\frac{Q^2 T_\infty}{C_p^2 \rho_\infty^2 g}\right)^{1/3}} = f\left(\frac{z}{z_0}\right), \tag{4.26}$$

and

$$\theta = \frac{\left(\frac{\Delta T_z}{T_\infty}\right) l_1^{5/3}}{\overset{*}{Q}{}^{2/3}} = f\left(\frac{z}{z_0}\right). \tag{4.27}$$

4.1.5 Flame Height Correlation

Several studies (Tang 2012; Lee 2009; Hu 2013; Klopovic 2001) have measured flame heights for fire emerging from a window opening and impinging on the façade wall. Experimentally, flame height is a visual observation. There is a consensus among fire researchers that the flame height is measured where flames are visible 50% of the times. This is called the mean flame height. More recently, Tang (2012) and Lee (2008) used a Charged Coupled Device (CCD) camera to obtain mean flame heights for their scaled experiments of façade fires. This provides a better quantitative establishment of flame heights.

Lee (2007) identified two domains for the mean flame height correlation as shown in Fig. 4.6.

The correlation takes the form

$$\frac{Z_f - Z_n}{l_1} = k\overset{*}{Q}_{\text{ex}}^{m}, \tag{4.28}$$

where Z_f is the flame height and Z_n is the height of the neutral pressure plane from the bottom of the window opening. In the experiments performed by Lee, the heat input $Q_{\text{in}} > Q_{\text{st}}$, and Lee defined $Q_{\text{ex}} = Q_{\text{in}} - Q_{\text{st}}$ (4.11). For small fires, $\overset{*}{Q}_{\text{ex}} < 1, m = 2/3$, and for large fires, $\overset{*}{Q}_{\text{ex}} > 1, m = 2/5$. The exponent $m = 2/3$ for flame height ratio has also been observed for experiments with wall fires (Hasemi 1984).

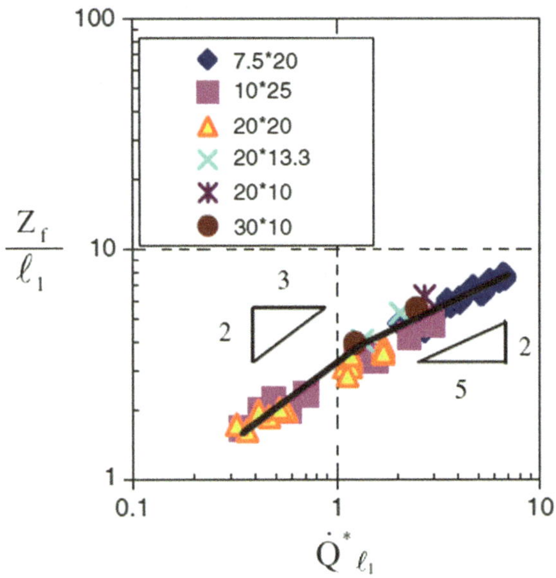

Fig. 4.6 Flame height correlation by Lee (2007). *Note* the legend in the figure refers to the window opening size, $W_V \times H_V$ in cm; the x-axis refers to $\overset{*}{Q}_{l_1}$ which is equivalent to $\overset{*}{Q}_{\text{ext}}$

The result from experiments performed by Tang (2012) on inert façades is given by (4.29) where (4.11) was used to calculate Q_{ex}, as

$$\frac{Z_f - Z_n}{l_1} = 2\overset{*}{Q}_{ex}^{0.44}. \tag{4.29}$$

For under-ventilated conditions, $Z_n \approx 0.35H_v$ (Yokoi 1960). It may be noted that the empirically derived exponent, 0.44, in (4.29) is close to the exponent $(2/5 = 0.40)$ reported in the literature for axisymmetric plume fires (Lee 2007).

4.1.6 Heat Flux

Thermal exposure from the flames emerging from the window opening can be characterized in terms of heat flux. This heat flux is a combination of convective and radiative energy from the hot gases and visible flames, given by

$$\dot{q}''_t = \dot{q}''_c + \dot{q}''_r, \tag{4.30}$$

where $\dot{q}''_t$, $\dot{q}''_c$, and $\dot{q}''_r$ are the total, convective, and radiative heat fluxes (W/m^2), respectively. Convective and radiative heat fluxes were introduced in Chap. 2 (see 2.4) and (2.5). Experimentally, the total heat flux may be measured using a Gordon or Schmidt–Boelter-type heat flux gauge.

Tang (2012) showed that for $\overset{*}{Q}_{ex} < 1.3$, fire ejected from the window opening behaves like a wall fire with air entrainment primarily from the front of the opening, whereas for $\overset{*}{Q}_{ex} > 1.3$, fire emerging from the window opening is axisymmetric. Hasemi (1984) developed a correlation for heat flux for wall fires using $\overset{*}{Q}^{2/3}$ as the scale for flame heights and maybe applicable to façade fires for $\overset{*}{Q}_{ex} < 1.3$. The correlation from Hasemi's work is reproduced in Fig. 4.7.

A seminal approach on façade fires was performed by Oleszkiewicz (1989, 1990) using large-scale test structures in a series of fire tests. One series included wood cribs as a fuel source, and the other series used propane gas burners. The propane gas fires enabled controlled heat input. The tests were conducted in a $5.95 \times 4.4 \times 2.75$ m fire test room. The tests used four different window opening sizes and fires ranging from 5.5 to 10.3 MW. The fire test room characteristics are summarized in Table 4.2.

The heat flux results are presented in Table 4.3 and have been numbered by the authors for discussion purposes. The measured heat fluxes at 0.5, 1.5, 2.5, and 3.5 m height above the window opening are presented. In the analysis of the data, $Q_{ex} = Q_{in} - Q_{fl}$ (4.5) was used, following Nishio (2018), with (4.1) for estimation of HRR at flashover.

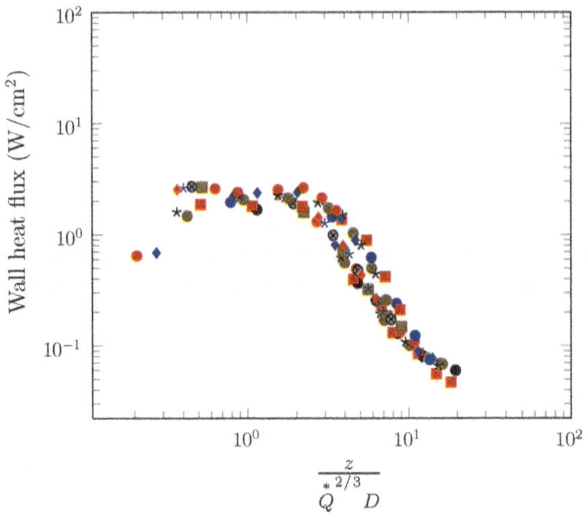

Fig. 4.7 Heat flux correlation (Hasemi 1984)

Table 4.2 Window openings used in NRC façade fire tests with propane burner tests for A_T = 109.3 m^2 (Oleszkiewicz 1989)

Window width m	Window height m	Window area, A_v m^2	$A_v\sqrt{H_v}$ m$^{3/2}$	l_1 (4.18) m	Window aspect ratio n	Q_{fl} (4.1) kW	Q_{st} (4.10) kW
0.94	2	1.88	2.66	1.48	0.94	1857	3990
0.9	2.7	2.54	4.17	1.77	0.7	2429	6255
2.6	1.37	3.56	4.17	1.77	3.8	2428	6255
2.6	2	5.2	7.35	2.22	2.6	3632	11,025
2.6	2.7	7.02	11.54	2.66	1.93	5213	17,310

The ventilation aspect ratio significantly affects heat flux. The heat fluxes are higher for larger values n. The location of heat flux gauges above the window was normalized with non-dimensional flame height similar to Hasemi's research. The data are presented in Fig. 4.8.

The window aspect ratio also has an influence on the heat transfer to the façade. The heat flux at 0.5 m location above the window appears to be offset toward the left as compared to data at the other three higher locations. Since this location is close to the window opening, it is more influenced by the momentum of the flames ejected from the window opening as it turns upward due to buoyancy forces. An example of this phenomenon is documented by Peng (2016) as shown in Fig. 4.3. This phenomenon is expected due to a lowering of the convective and radiant heat transfer to the façade closer to the window opening.

Table 4.3 Data of heat flux from NRC façade tests with propane burners (Oleszkiewicz 1989)

#	W_v	H_v	n	Q_{in}	Q_{fl}	Q_{ex}	Q^*_{ex}	Heat flux (kW/m^2) measured at			
	m	m		MW	kW	kW	kW	0.5 m	1.5 m	2.5 m	3.5 m
1	0.94	2	0.94	5.5	1857	3643	1.232	43.9	12.4	7.7	3.9
2	0.94	2.7	0.7	5.5	2429	3071	0.662	19.2	6.3	3.5	1.7
3	2.6	1.37	3.8	5.5	2428	3072	0.663	24.5	22.9	13.2	11.5
4	2.6	2	2.6	5.5	3632	1868	0.228	10.5	5.2	4.5	2.9
5	2.6	2.7	1.93	5.5	5213	287	0.022	6.5	2.9	2	1.4
6	0.94	2	0.94	6.9	1857	5043	1.706	58.6	17.7	9.9	5.1
7	0.94	2.7	0.7	6.9	2429	4471	0.964	34.8	10.4	6	2
8	2.6	1.37	3.8	6.9	2428	4472	0.965	53.2	33.1	17.2	15.6
9	2.6	2	2.6	6.9	3632	3268	0.4	17.4	9.4	7.4	5.4
10	2.6	2.7	1.93	6.9	5213	1687	0.132	11.4	5.3	4.2	2.9
11	0.94	2	0.94	8.6	1857	6743	2.281	75.5	25.9	15.9	8.1
12	0.94	2.7	0.7	8.6	2429	6171	1.331	53.2	15.9	9.8	4.8
13	2.6	1.37	3.8	8.6	2428	6172	1.331	104.3	58.6	51.2	28.3
14	2.6	2	2.6	8.6	3632	4968	0.608	29.5	14.8	12.6	8.2
15	2.6	2.7	1.93	8.6	5213	3387	0.264	17.4	8.1	5.7	3.6
16	0.94	2	0.94	10.3	1857	8443	2.856	NM	NM	NM	NM
17	0.94	2.7	0.7	10.3	2429	7871	1.698	68.3	23.2	13.7	6.7
18	2.6	1.37	3.8	10.3	2428	7872	1.698	208.7	122.4	103.9	56.5
19	2.6	2	2.6	10.3	3632	6668	0.815	43.4	20.8	16.3	9.6
20	2.6	2.7	1.93	10.3	5213	5087	0.397	29.1	12.8	9.1	5.6

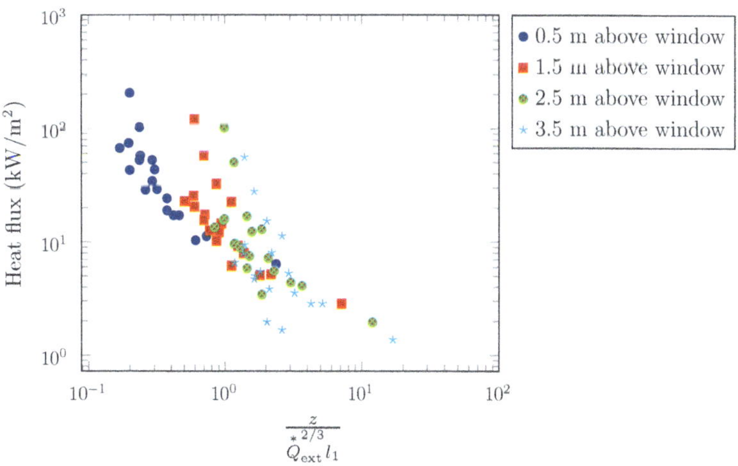

Fig. 4.8 Heat flux on façade from NRC tests

The window aspect ratio also has an influence on the heat transfer to the façade. The heat flux at 0.5 m location above the window appears to be offset toward the left as compared to data at the other three higher locations. Since this location is close to the window opening, it is more influenced by the momentum of the flames ejected from the window opening and as it turns upward due to buoyancy forces before it attaches to the façade wall. This phenomenon was photographed by Peng (2016) as shown in Fig. 4.3. The window aspect ratios in Peng's experiments are in the same range as the NRC data. The smaller window opening factor results in higher flame momentum from the window and lowers the convective and radiant heat transfer to the façade closer to the window opening.

An example to demonstrate influence of the window aspect ratio factors is tests 17 and 18 for the 10.3 MW fire. These tests have identical ventilation factor ($A_v\sqrt{H_v}$) and heat release but have a very different heat flux exposure on the façade. In test 18, the flames are expected to be closer to the wall ($n = 3.80$) versus test 17 where $n = 0.7$. Similar influence may be observed for tests 2 and 3 (5.5 MW fire); tests 7 and 8 (6.9 MW fire); and tests 12 and 13 (8.9 MW fire).

Heat flux exposure data from several façade experiments have been complied. Two sources include the measurements in the JIS A1310 standard (Nishio 2018) and experiments by Peng (2016). The data by Peng were digitized from the graph in the paper. The combined NRC and these data are presented in Fig. 4.9. All these data follow similar trend with the scaling factor for vertical distance from the window opening.

Lee (2007) has attempted to include the influence of the window scales in his correlation for heat flux. The correlation is presented in Fig. 4.10 where the heat flux was attenuated for the window opening parameters (i.e., $e^{(0.6(H_v/l_1))}$). It may be observed that the x-axis, representing distance above the window opening, is scaled

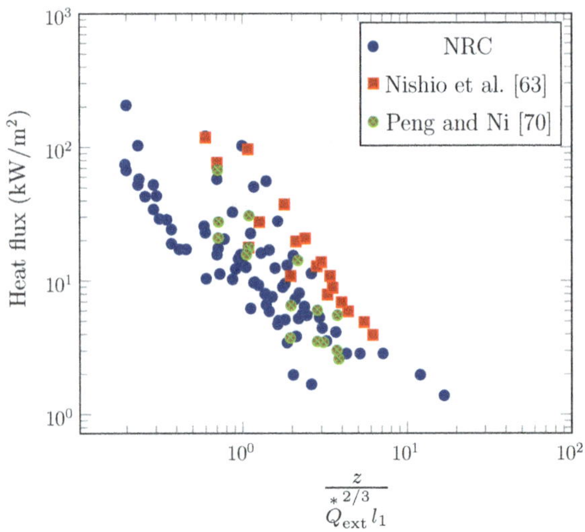

Fig. 4.9 Comparison of NRC heat flux data with other façade fire experiments

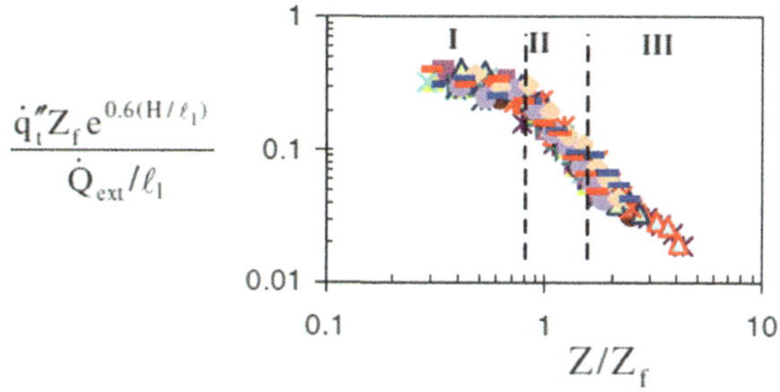

Fig. 4.10 Heat flux correlation by Lee (2007)

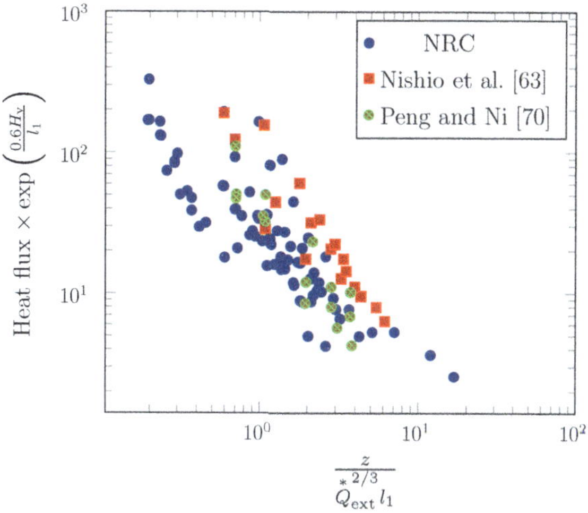

Fig. 4.11 Heat flux data with Lee's attenuation factor

with flame height (Z_f). The three regions identified in Fig. 4.10 represent regions for solid flame (I), intermittent flaming (II), and the convective thermal plume (III).

Following Lee's approach, the heat flux data from NRC, JIS A 1310 Nishio (2018) and Peng (2016) were re-analyzed by attenuating the heat flux with for the window parameters. The results are shown in Fig. 4.11; the attenuation factor appears to account for the influence of the window parameters on the heat flux on the façade.

4.2 Fire Performance of Testing of Façade Assemblies

Many countries consider a room and content fire that breaks out from the room of origin as an important scenario for standard testing. In this scenario, the emerging flames and hot gases thermally impact the façade material. Some countries (e.g., Germany) have focused on fire originating on the exterior of the façade, for example, by trash or debris. Some countries (e.g., India) are considering developing code requirements and identifying test parameters for façade systems that may be appropriate to their experience.

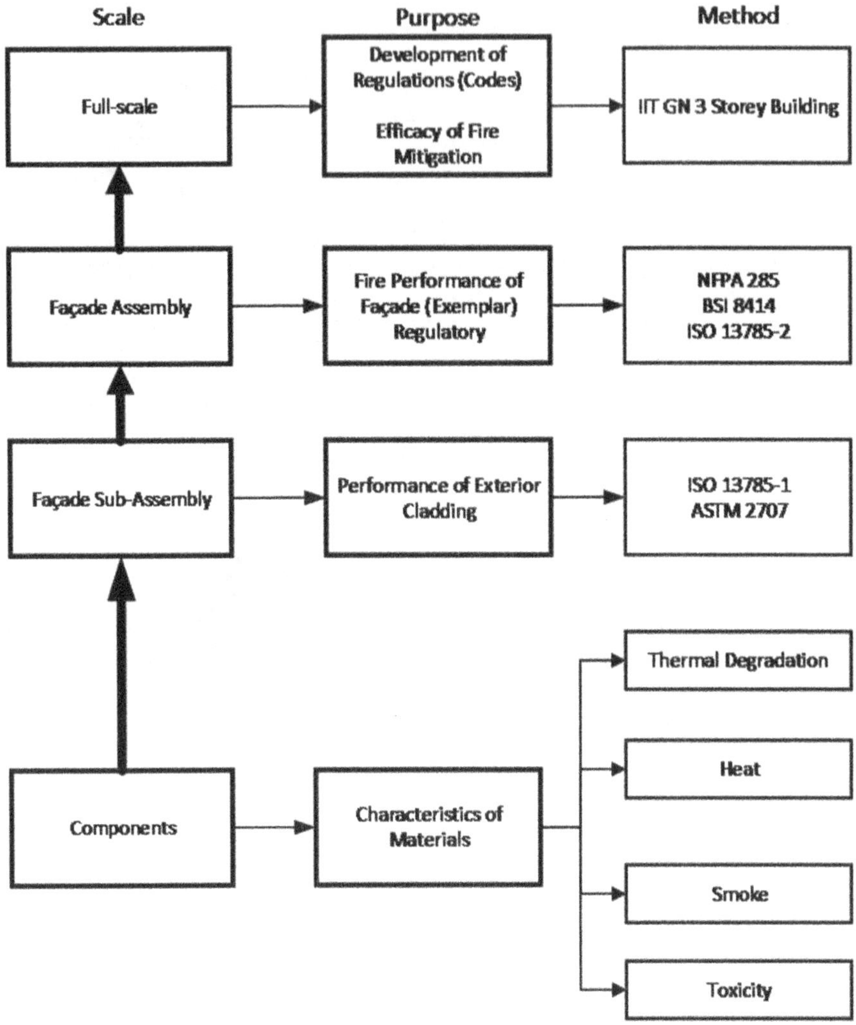

Fig. 4.12 Fire performance testing of façade systems

Figure 4.12 summarizes the different scales of façade testing, their purposes, and example standard test methods.

4.2.1 Full-Scale Testing

The objective of full scale of testing is to develop data to support code development as well as to demonstrate the efficacy of different fire mitigation strategies. The test facility should have the capability to introduce both passive and active fire mitigation strategies that provide time for occupants to egress safely as well as allow the fire service time to bring a fire under control and suppress it. Based upon fire experience involving façades in high-rise buildings, a full-scale test facility should be able to investigate the following:

- Vertical and lateral flame propagation over the exterior face of the wall assembly;
- Vertical flame propagation within the combustible core, air cavities, or within combustible components from one story to one above;
- Vertical flame propagation over the interior surface of the wall assembly from one story to the next;
- Lateral flame propagation from the compartment of fire origin to adjacent compartments or spaces;
- Interaction between façade and compartment fires at upper stories; and
- Effects of wind conditions on fire behavior.

Such a test facility has been developed at the Indian Institute of Technology Gandhinagar to facilitate fire safety regulations in India. The test fixture is a three-story (G+2) steel–concrete structure with floors protected from direct fire exposure. Three sides of the buildings may be used for installing façade systems for test purposes. A before and after test view of this facility is shown in Fig. 4.13, and a detailed discussion on its design features is presented in Sect. 5.1.

The facility has been used for a number of research projects (Gandhi 2017; Srivastava 2018, 2020) to investigate fires with façade systems com-

(a) Three-story façade test structure. (b) After a full-scale test.

Fig. 4.13 Full-scale test facility at IIT Gandhinagar

monly used in India. The full-scale facility has also been used to investigate other fire protection passive concerns with high-rise buildings such as perimeter protection, fire containment with fire doors, and fire stopping of through-penetration (e.g., plumbing pipes and electric wiring conduits). It is equipped to add active fire suppression with automatic sprinkler systems and a smoke/gas collection system from different compartments as per the needs.

Based upon the experience in a country, the building regulations may develop fire safety objectives for façades that include the following fire performance metrics:

- Flame spread on the façade (vertical and horizontal);
- Fire spread from the room of fire origin to the room above;
- Perimeter protection;
- Detailing around the window opening (e.g., architectural obstructions);
- Smoke;
- Heat transmission to the façade (temperature and heat flux);
- Fire spread through gaps in the façade system;
- Posttest damage assessment of façade and other building components.

These metrics also assist in designing the test and the measurement system used for façade assembly.

4.2.2 Façade Assembly Testing

Façade assemblies are composites of different materials, each with a unique function. For this reason, for regulatory purposes, façades are tested in a large-scale test where a façade assembly example is installed according to the manufacturer's instructions. These instructions include methods for fastening and perimeter fire protection (e.g., gaps between the wall and the floor). Country and region-specific regulations require testing façade systems to meet the defined fire protection objectives. For example, in the USA, the façade systems are tested using NFPA 285; in the UK, they are tested using BS 8414-1 or BS 8414-2; in the EU region, there are several different tests that include LEPIR-2, SP Fire 105, and ISO 13785-2. These tests may be considered assembly-scale tests where the façade system incorporates all the essential features of the end-product, including procedures for installation. A list of test methods across North America and Europe (Development of a European Approach 2018) is presented in Table 4.4.

Figure 4.14 shows the geometric configurations for some of the aforementioned tests where photographs and images were available in the literature. These tests may be categorized as either a fire exposure to the façade from fires emerging from an opening or a fire exposure directly on the exterior of the façade wall. Some of the tests have a single façade wall (e.g., NFPA 285, SP 105, LEPIR-2), whereas some have an enclosed corner (wing) as in BS 8414 and ISO 13785-2. The scale of the tests has implications for the size of the façade tested and the installation details. It may be noted that DIN 4201-20, while being a medium-scale test, represents a

Table 4.4 List of test methods

Country	Test method	Scale	Fire scenario
Canada	CAN/ULC-S134-92	Large	Fire inside the building and flames emerging from the opening to thefaçade
Czech Republic	ISO 13785-1:2002	Medium	Fire exposure on external to the façade
Finland	Engineering guidance 16 (unofficial test method)	Large	Fire inside the building and flames emerging from the opening to the façade
France	LEPIR-2	Large	Fire inside the building and flames emerging from the opening to the façade
Germany	Technical regulation A 2.2.1.5	Large	Fire exposure on external to the façade
Hungary	MSZ 14800-6:2009	Large	Fire inside the building and flames emerging from the opening to the façade
Poland	PN-B-02867:2013	Medium	Fire exposure on external to the façade
Switzerland, Lichtenstein	Prüfbestimmung für Aussenwandbeklei-dungssysteme	Large	Fire inside the building and flames emerging from the opening to the façade
Slovakia	ISO 13785-2:2002	Large	Fire inside the building and flames emerging from the opening to the façade
Sweden, Norway, Denmark	SP Fire 105	Large	Fire inside the building and flames emerging from the opening to the façade
Switzerland, Austria	ÖNORM B 3800-5	Medium	Fire inside the building and flames emerging from the opening to the façade
Switzerland, Germany	DIN 4102-20	Medium	Fire inside the building and flames emerging from the opening to the façade
UK, Republic of Ireland	BS 8414-1:2015 and BS 8414-2:2015	Large	Fire inside the building and flames emerging from the opening to the façade
USA	NFPA 285	Large	Fire inside the building and flames emerging from the opening to the façade

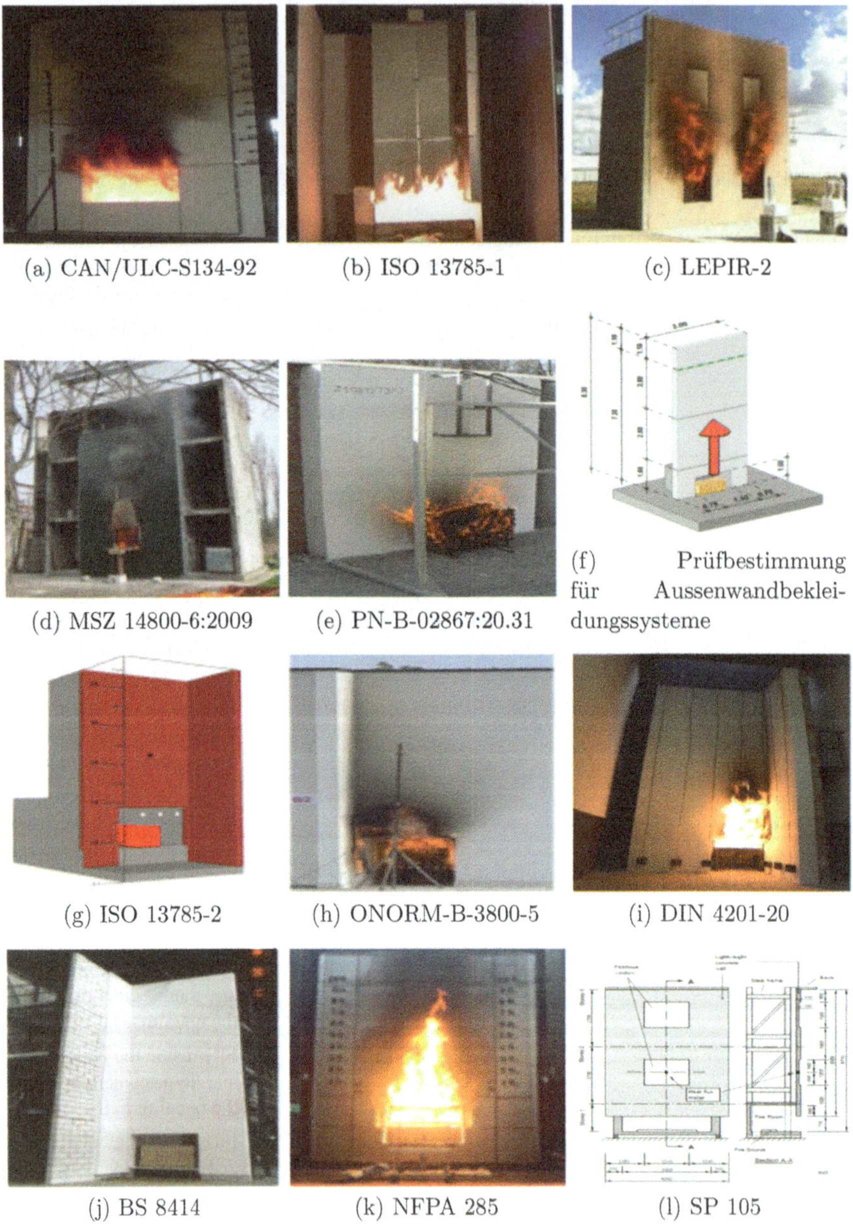

(a) CAN/ULC-S134-92 (b) ISO 13785-1 (c) LEPIR-2

(d) MSZ 14800-6:2009 (e) PN-B-02867:20.31 (f) Prüfbestimmung für Aussenwandbekleidungssysteme

(g) ISO 13785-2 (h) ONORM-B-3800-5 (i) DIN 4201-20

(j) BS 8414 (k) NFPA 285 (l) SP 105

Fig. 4.14 Fire test configurations

Table 4.5 Characterization of façade tests with fire emerging from an opening

Test method	$W_v \times H_v$ m	Fire source	n	$A_v\sqrt{H_v}$ $m^{3/2}$	l_1 m	l_2 m	Q_{fl} (4.1) kW	Q_{st} kW	Max. Q_{ex} kW	Max. Q^*_{ex}
DIN 4102-20	1 × 1	320 kW gas fire or 30 kg wood crib	2	1	0.5[a]	NA	NA	NA	320	0.51
ISO 13785-2	2 × 1	Propane gas, Fig. 4.16	4	2	1.32	1.19	1255	2500	1130	1.82
BS 8414	2 × 2	400 kg wood crib	2	5.66	1.00[a]	NA	NA	NA	3000	2.71
LEPIR-2	1 × 1.5[b]	2X 300 kg wood cribs	1.33	1.57	1.2	1.36	1058[c]	1967	958[c]	0.77[c]
NFPA 285	1.98 × 0.76	2-Natural gas burners, Fig. 4.17	5.2	1.31	1.11	0.97	832	1640	760[d]	0.32

Notes [a]Scale factor l_1 for BS 8414 and DIN 4102-20 estimated the width from video documentation of flames emerging from the combustion chamber
[b]LEPIR has two window openings
[c]Each window opening
[d]Based on value of the window burner at the end of 30-min duration NA: Not applicable as combustion chamber size is comparable to the fuel source

scenario of fire emerging from a window opening, whereas PN-B-02867:2013 is a direct fire exposure from flames to the façade surface.

Some of the tests with fires emerging from an opening are characterized in Table 4.5 relative to window opening scales and HRR for flashover, stoichiometric combustion of incoming air, and heat release of excess unburnt fuel outside the test structure. NFPA 285 test is unique among the large-scale tests because it has a second gas burner at the top of the window opening that is initiated five minutes after the room burner.

Both DIN 4201-20 and BS 8414 tests have relatively small combustion chambers, and thus the flames from the fuel sources emerge from the opening and impinge directly on the façade system like a wall fire exposure. A photograph of the BS 8414 depicts this phenomenon as shown in Fig. 4.15. This indicates the flame emerging from the window is like wall flames (see Fig. 4.3) and the flame height and heat flux may be represented by correlations developed by Hasemi (1984).

For these two tests, HRR for flashover was not calculated as they do not represent a traditional room and content fire. The HRR from excess fuel may be estimated to be that from the fuel sources used.

(a) Combustion chamber and positioning of wood crib.

(b) Flames from wood crib 1 minute after ignition.

Fig. 4.15 BS 8414 façade fire test

DIN 4102-20 and BS 8414 tests are expected to be wall fires with small value of scale factor l_2. NFPA 285 has the highest window aspect ratio and length scale, l_2, that correlates to the momentum of gases emerging from the window. The length scale, l_2, also indicates that the ISO 13785-2 and NFPA 285 will have flames closer to the façade than the LEPIR-2 test.

To obtain a comparison of the fire exposure to the façade in these large-scale tests, data on HRRs under well-ventilated conditions from the fuel sources used are needed. The HRR from the wood cribs used in LEPIR-2 has been reported (Dréan 2008) to have a peak of 2400 kW each. This indicates that there is combustion of unburned fuel from the crib decomposition outside the window opening.

ISO 13785-2 has a graduated propane gas flow over a period of 25 minutes with 15 min of flow at 120 g/s corresponding to a heat release of 5280 kW as shown in Fig. 4.16. NFPA 285 test has a room burner and a window line burner with graduated HRR over the test duration of 30 min as shown in Fig. 4.17. The heat input of the NFPA 285 room burner at 30 min (904 kW) is over the estimate for HRR for flashover (760 kW). The window burner supplements the room burner to simulate gases burning outside the window starting 5 min after the start of the test.

The HRR data were used to estimate the non-dimensional fire strength, $\overset{*}{Q}_{ex}$ (4.11) for the different fire tests as shown in Table 4.5. BS 8414 has the highest $\overset{*}{Q}_{ex}$ value indicating a higher flame height and radiant exposure from the flames to the façade cladding.

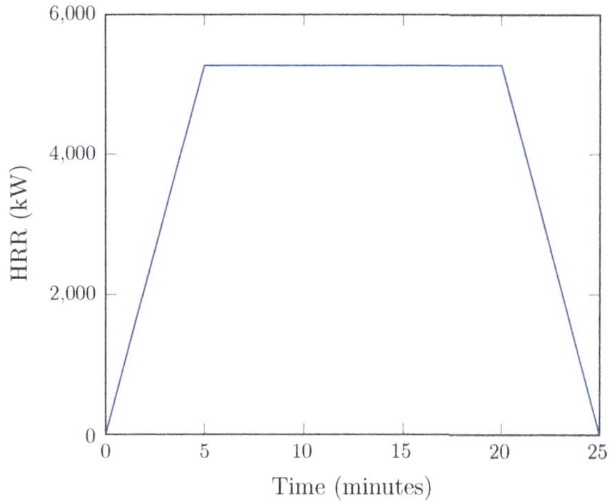

Fig. 4.16 ISO 13785-2 HRR

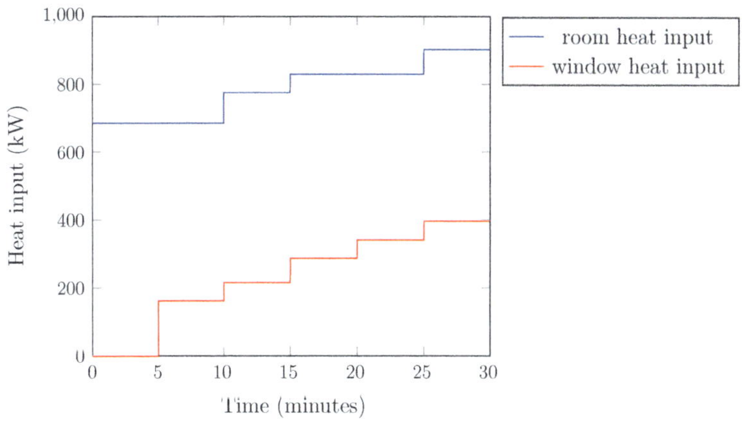

Fig. 4.17 NFPA 285 heat input

The thermal exposure to the façade from the fire is measured, as part of their calibration procedures, using non-combustible façade wall materials on several of the assembly-scale tests. This is presented in Table 4.6. Since the fire exposure in NFPA 285 is controlled (Fig. 4.18) in the combustion room as well as the window, the standard provides temperature and heat flux on the façade material during the calibration procedures.

The discussion thus far has highlighted that countries seek to regulate installation of façades with combustible materials through fire testing of a finished façade system under representative installation conditions. Since the regulatory assembly-scale tests can be expensive, subassembly- and component-level testing can facilitate design and selection of component materials. The configuration and conditions of

Table 4.6 Comparison of heat flux on façade from fire exposure

Test method	Fire exposure	Temperature	Heat flux kW/m^2
DIN 4102-20	Gas burner turned off at 20 min, and wood crib extinguished after 30 min; followed by 40-min posttest observation	Not specified	60 at 0.5 m, 35 at 1 m, 25 at 1.5 m
ISO 13785-2	Fire exposure adjusted to provide gradual fire intensity increase 4–6 min before and gradual decrease 4–6 min after for a total test duration of 23–27 min	Greater than 600 °C directly above the window opening	Average (over 15 min) after flames emerge from the opening: 55 at 600 mm; 35 at 1.6 m
BS 8414	Fire exposure for 30 min followed by 30 min of posttest observation	600 °C directly above the opening for 20 min	Maximum heat flux of 70 at 1m above opening
LEPIR-2	30 min fire exposure	Temperatures measured 150 mm from the surface average of 500 °C; peak of 800 °C above the opening	Not specified
NFPA 285	30 min total fire exposure with window burner starting at 5 min	Temperature measured at 1 ft intervals above the window (Fig. 4.18 shows temperature at 0.3, 0.91, and 2.1 m above opening)	Measured at 2-ft, 3-ft, and 4-ft above the window (Fig. 4.18 shows heat flux at 0.91m (3 ft) above opening)

testing, we have seen, may vary but have similar features providing high heat flux and temperature impinging upon the façade materials. To this end, subassembly- and material-level tests may be performed and analyzed using the framework shown in Fig. 4.12.

4.2.3 Subassembly-Level Testing

The purpose of the façade subassembly test is to develop data on performance of the composite façade when exposed to a fire source with equivalent severity as used in façade assembly test. Table 4.6 provides guidance for the severity of exposure required for country-specific regulatory test. Three subassembly tests discussed in this chapter are:

Fig. 4.18 NFPA 285: Temperature and heat flux on the façade from heat input

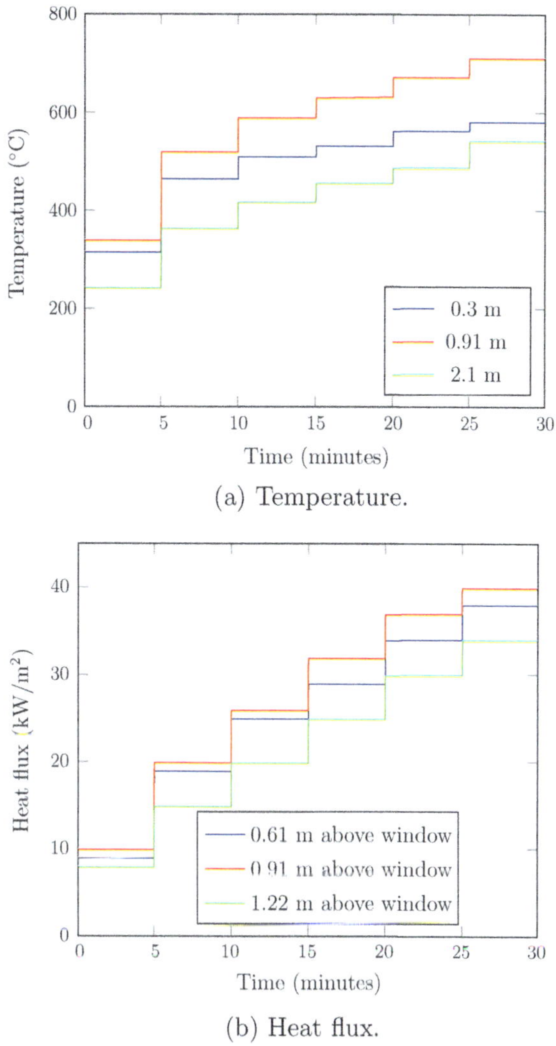

- ISO 13785-1: Reaction to Fire Tests for Façades Part 1: Intermediate-Scale Test;
- ASTM E2707 Standard Test Method for Determining Fire Penetration of Exterior Wall Assemblies Using a Direct Flame Impingement Exposure;
- JIS A 1310 Test Method for Fire Propagation Over Building Façades.

4.2.3.1 ISO 13785-1

The test consists of two test panels forming an included right angle as shown in Fig. 4.19. The back panel is 2.4×1.2 m wide, and the side panel is 2.4×0.6 m wide. A propane gas burner (1.2×0.1 m) provides a 100 kW fire exposure to the panels.

Using (4.21), the non-dimensional strength may be calculated, using burner width as the flame characteristic dimension ($l_1 = 1.2$ m), as

$$\overset{*}{Q}_{\text{ex}} = \frac{100}{\rho_\infty T_\infty C_{\text{p}} \sqrt{g} l_1^{2.5}} = 0.057. \tag{4.31}$$

Using (4.29) yields a flame height estimate of 0.3 m. Hasemi's equation for wall flames (Hasemi 1984) yields a flame height estimate of 0.2 m. This is in alignment with visual observations (Dréan 2019) shown in Fig. 4.20 from a photograph 1 min after ignition. The non-dimensional strength indicates a relatively lower intensity fire exposure with smaller heat flux to the façade test samples as compared to the assembly-scale regulatory tests.

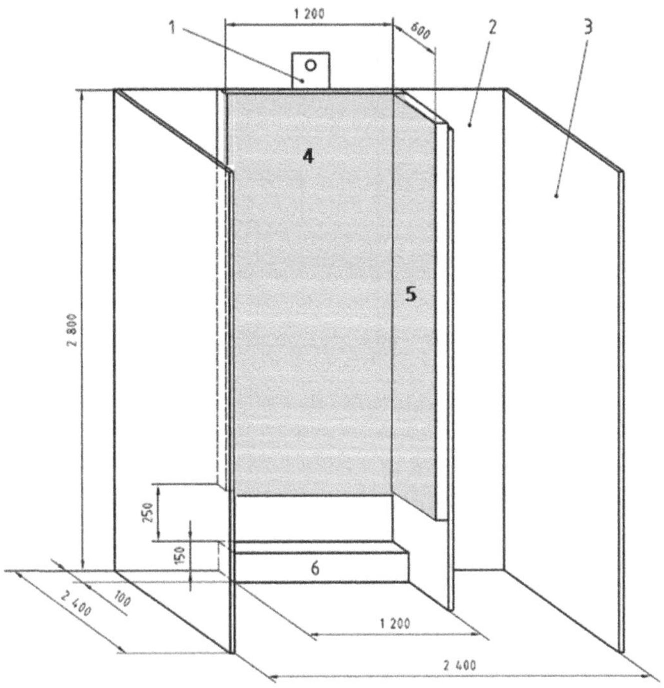

1. Heat flux gauge; 2. Back Panel; 3. Draft free panel;
4. Test Specimen back panel; 5. Test Specimen side panel; 6. Gas burner

Fig. 4.19 ISO 13785-1 intermediate-scale test for façades

Fig. 4.20 Photograph of flames from ISO 13785-1 test, 1 min after ignition

Limited testing results with this test for façade panels are available (Guillaume 2018) for 3 different façade systems and 3 insulation types. One of the systems (ACP with PE) is similar to that used in Grenfell Tower. The test of this system resulted in a HRR exceeding 5 MW. There are not sufficient data to demonstrate if this intermediate-scale test can correlate to large-scale regulatory tests and more work is required in this area.

4.2.3.2 ASTM E2707

The test was developed for assessing the performance of cladding materials exposed to external fires such as wildland fires. However, it has been used in the USA as a screening test of façade designs prior to performing the NFPA 285 test required by US fire regulations. A schematic of the test setup is shown in Fig. 4.21. The intended purpose of the test method is to measure the ability of the test sample to resist fire penetration from the exterior into the wall cavity or unexposed side of the test assembly under the conditions of exposure.

In the test, a vertical test sample (2.44×1.22 m wide) is exposed to a gas burner fire diffusion flame of 150 kW. The burner is 1×0.1 m wide. The test duration is 10 min. Using (4.21), the non-dimensional strength may be calculated, using burner width as the flame characteristic dimension ($l_1 = 1.0\,\mathrm{m}$), as

$$\overset{*}{Q}_{\mathrm{ex}} = \frac{150}{\rho_\infty T_\infty C_\mathrm{p} \sqrt{g} l_1^{2.5}} = 0.136. \tag{4.32}$$

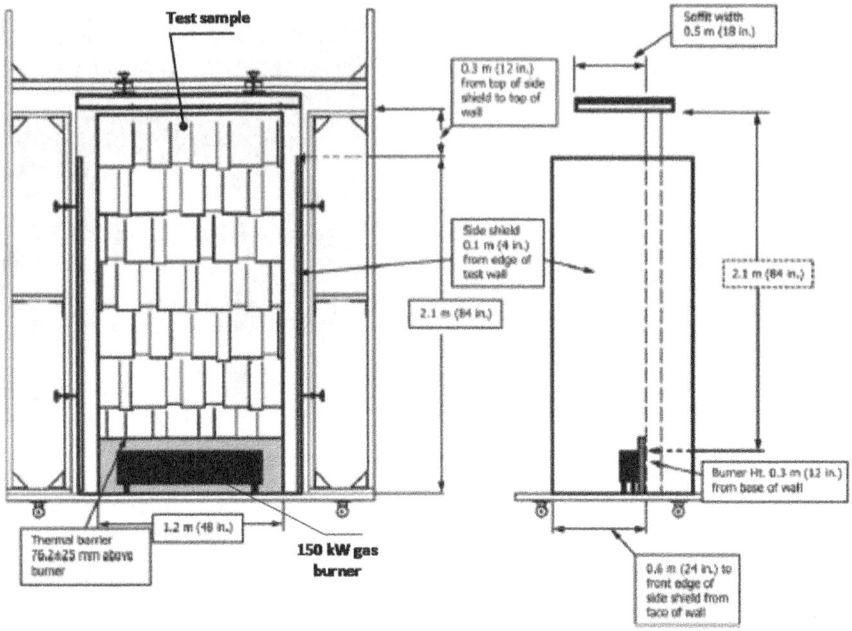

Fig. 4.21 ASTM E2707 test setup

The exposure provided by the ASTM 2707 is higher than the ISO 13785-1 test but still lower than any of the regulatory tests.

4.2.3.3 JIS A 1310

In Japan, an intermediate test, JIS A1310:2015, has been developed (JIS 2015) for regulating façade systems. The test method simulates a small-scale room and content fire with flames emerging from a combustion chamber onto the test façade. Yoshioka (2016) has reported the parameters of the test and results relative to several façade systems. The test setup is shown in Fig. 4.22. The test sample in this test is of size 1.82×2.73 m (width $\times$ height), with a square window opening of 0.91×0.91 m. The test uses a propane gas burner. The characteristics of this test in comparison with some of the regulatory tests are shown in Table 4.7.

The heat flux on the façade at heights above the top of the window opening with four heat input levels is shown in Fig. 4.23. The x-axis of the plotted data is the non-dimensional distance ($z/Q_{ex}^{*2/3} l_1$). The 340 kW fire is close to the HRR estimate for flashover in the JIS test, and thus, the plume outside the window is expected to be a thermal plume. For comparison, heat flux data from large-scale regulatory tests are also shown in Fig. 4.23. The heat flux data from the regulatory tests, plotted with

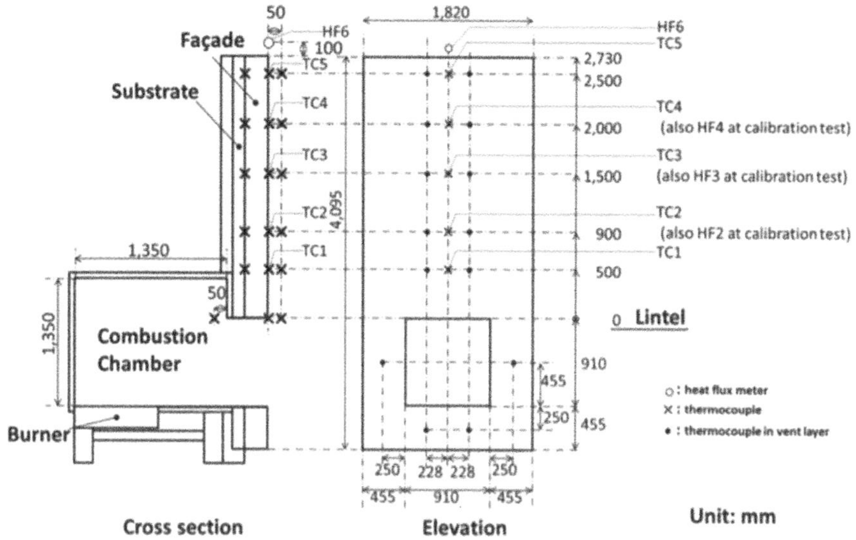

Fig. 4.22 JIS A1310:2015 test setup

Table 4.7 Comparison of heat flux on façade from fire exposure

Test method	W_v (m)	H_v (m)	n	$A_v\sqrt{H_v}(m^{3/2})$	l_1 (m)	Q_{fl} (kW)	Max. $\dot{Q}^*_{ex}$
JIS A1310	0.91	0.91	2	0.79	0.91	328	0.014 (340 kW); 0.31 (600 kW); 0.65 (900 kW)
DIN 4102-20	1	1	2	1	0.5	NA	0.51
ISO 13785-2	2	1	4	2	1.32	1255	1.82
BS 8414	2	2	2	5.66	1.00	NA	2.71
LEPIR-2	1	1.5	1.3	1.57	1.2	1058	0.77
NFPA 285	1.96	0.78	5.2	1.31	1.11	832	0.32

Notes NA—Not applicable due to small size of the combustion chamber

non-dimensional height above the window opening, are within the range of data from the JIS tests indicating similar physical fire plume phenomenon.

The JIS A1310 has the advantage that it can represent heat flux exposure of different regulatory tests by varying the heat input, Q_{in}. For example, a heat input of 940 kW is expected to provide heat flux from the fire exposure to the façade in the same range as the ISO 13785-2 and DIN 4102-20 tests.

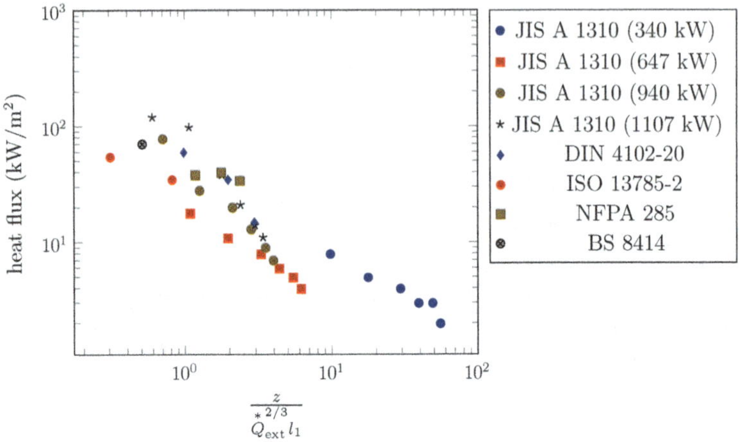

Fig. 4.23 JIS A 1310 heat flux comparison with regulatory tests

4.2.4 Component-Level Tests

The objective of the component-level testing is to enable selection of materials used in the façade system based upon the required design and fire performance. These data may be available in the form of material suppliers' datasheets. The component-level data provide important inferences on performance of the system. The data may also be used as input for computer simulations as discussed in the next section.

Since façade assemblies are composites and must be evaluated as a single system, the scope of the materials and component-level performance is limited to the material's selection phase. Because the fire performance of a façade assembly entails interaction of multiple components during the fire exposure, a façade assembly may yield a different fire performance response than any individual component or material.

The key considerations in the design of a façade are reiterated in Table 1.1. The selected component properties have to meet the functional needs of the façade with additional constraints of fire safety, cost, installation ease, and aesthetics.

The data relevant for flammability include (i) polymer characterization; (ii) thermal performance; (iii) heat; (iv) smoke; and (v) toxicity that are identified in Fig. 4.24; some of these are discussed in Chap. 3. A list of component-level tests is presented in Table 4.8. Materials with the potential for direct exposure to fire require better fire performance.

The testing and results of components and materials typical in façade systems are presented by McGrattan (2019). Additional properties such as density, thermal conductivity, and specific heat are important for developing a phenomenological model for heat transfer to the materials. For metal, exterior cladding, thermal expansion, and strength properties may also be required.

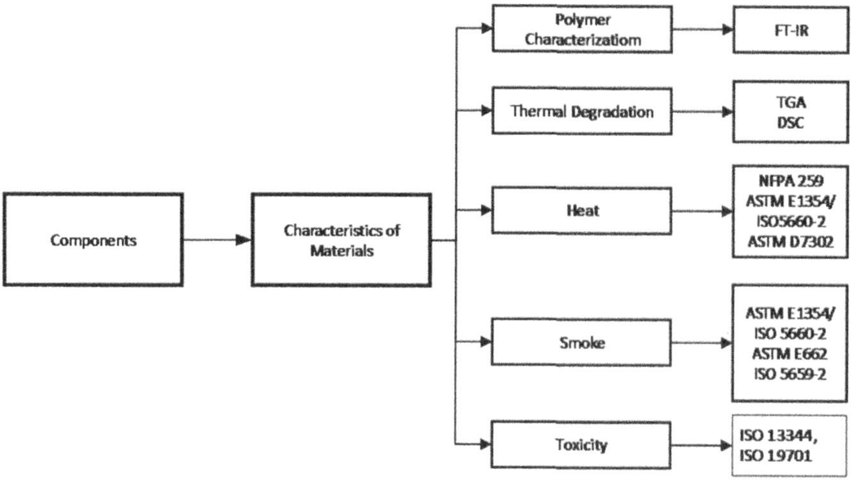

Fig. 4.24 Component-level testing

4.3 Computer Simulation

The assembly-scale tests used for regulatory purposes evaluate not only the façade system but also the installation procedures developed by the manufacturers. However, the results are constrained by the parameters of the test. For example, each of the assembly-scale tests has their unique window openings, and the façade systems tested are flat surfaces. Thus, while the results provide important evaluation of performance with exposure to flames and heat from a fire, the systems tested do not represent actual architectural designs of a buildings. One of the features of the assembly-scale tests is that window aspect ratio (l_2/l_1) is relatively low, implying a 'wall-like' fire exposure and thus potentially higher heat flux to the façade exterior (ref. NRC experiments).

Computer simulations can be used to understand the influence of architectural features on fire exposure to the façades (i.e., temperature and heat flux). The results may then be compared with the standard test to evaluate how the architectural features mitigate the fire exposure. It is recommended that guidelines provided by the SFPE G.06 (2011), AIAA (2011), and ISO 16730-1 (2015) be referenced when using computer-based simulation.

Fire Dynamics Simulator (McGrattan, 2014a) appears to be the Computational Fluid Dynamics (CFD) model of choice for researcher modeling façade fires. McGrattan (2014b) provided a discussion on sensitivity of various metrics to input uncertainty. Anderson (2017), using Fire Dynamics Simulator (FDS), provided an excellent summary of modeling of SP 105, BS 8414, and ISO 13785-2 fire tests. The paper also discussed a novel approach using statistical methods to investigate how uncertainty in input parameters may influence modeling results. It is a robust method that combined statistical methodology with a deterministic model to develop a range of results in comparison with experiments. Dréan (2008) modeled fire exposure of

Table 4.8 Component-level tests

Property	Test method	Comment
Polymer characterization		
Polymer type, chemistry	FT-IR (ASTM E1252), EDXRF (ASTM F2617, D6247)	Identifies the polymer chemistry and functional groups
Thermal degradation		
Thermal decomposition rate	TGA (ASTM E1311/ ISO 11358)	Measures mass loss as a function of time in either N_2 environment of air. Data used to develop understanding thermal response of material to elevated temperature; may be used for decomposition kinetics. The test uses 15–30 mg sample
Melting temperature	DSC (ASTM D794-06)	Measures melting temperature for polymeric materials used in the façade. The test uses 15–30 mg sample
Heat		
Heat of combustion	Bomb calorimeter NFPA 259/ISO 1716	Measures gross heat of combustion in an enriched oxygen and elevated pressure environment. The test uses 0.5 g sample
Ignition response	ASTM E1354/ISO 5660-2	Measure ignition time under a range of heat flux conditions. The test uses a 100 × 100 mm sample
Heat release	Microcombustion calorimeter (ASTM D7302)	Determines heat generated under ramped heating conditions. The test uses 5–10 mg sample
Heat, smoke, and burning		
Heat release and smoke release rates	ASTM E1354/ISO 5660-2	Measures ignition time, HRR, smoke release rate, heat of combustion, smoke extinction area under controlled heat flux condition. The test uses a 100 × 100 mm sample
Flame spread	ASTM E1321	Measures lateral ignition and flame spread—conditions to sustain flame spread after ignition
Smoke		
Smoke density	ASTM E662/ISO 5659-2	Measures smoke density generated under smoldering and flaming conditions. The test uses a 75 × 75 mm sample
Toxicity		
Fractional effective dose/LC50	ISO TS 19700 and ISO 19706; ISO 13344 and ISO 19701	The tests provide methodology for gas sampling and analysis to determine the lethality of combustion products

the LEPIR-2 façade test. The calculated radiative heat flux 4 m from the window opening (a specification of LEPIR-2 test) was within 20% of the experimental value. These two papers are good resources for the fire load input as well as the setup of the modeling parameters. The papers demonstrate that CFD codes such as FDS can be used to determine the fire exposure condition for a façade fire with flames ejected from the window opening.

Zhao (2016) used FDS to investigate the influence on window opening on façade fires. The paper provided a good reference on sensitivity analysis relative to grid size. It also highlighted inferences that need to be developed to compare with experimental results. For example, it discussed how to infer flame heights from FDS results based upon temperature and Heat Release Rate Per unit Volume (HRRPUV).

The ability of FDS to model fire exposure to façade opens several opportunities to use the model as a tool to assess fire hazards relative to the standardized test. These include (i) influence of façade system design and (ii) façade fire performance.

4.3.1 *Influence of Façade System Design*

The influence of window aspect ratio on the fire plume trajectory out of the window has been discussed previously. Other features such as projections from balconies or wings can influence fire exposure to the façade spandrel. Some examples of these features are shown in Fig. 4.25.

These architectural features around the window opening can influence fire plume trajectory once ejected from the window opening. For example, horizontal projections above a window can deflect fire plumes away from the façade surface and reduce heat flux and temperature above the window opening, whereas wings may reduce air entrainment and increase the fire exposure. The effect of horizontal projection was also investigated by Oleszkiewicz (1991) who showed that a 1 m horizontal projection reduced heat flux 1 m above the window by 85%. Peng (2016) provides

(a) Balcony projection. (b) Window projection. (c) Wings.

Fig. 4.25 Building architectural features that may influence façade fire plumes

(a) No obstruction.

(b) 0.5 m obstruction.

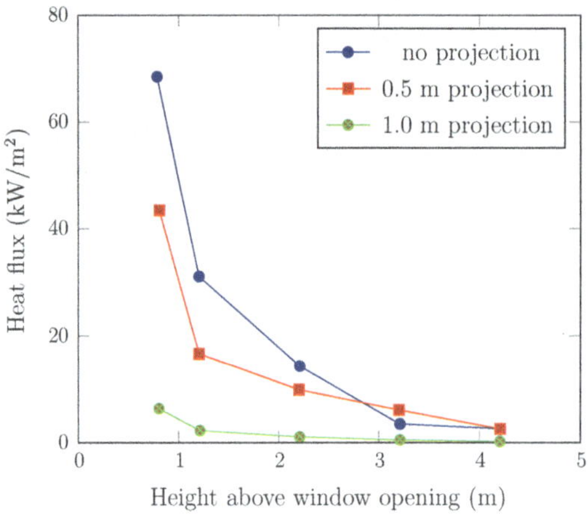

(c) Heat flux.

Fig. 4.26 Influence of projection above the window opening on ejected fire plume

an excellent visual documentation of the projections deflecting the fire plume away from the façade surface above them as shown in Fig. 4.26.

The utility of CFD simulations to explore the influence of architectural features is exemplified by a demonstration conducted in FDS with a model of the JIS A1310 test (JIS 2015) by the authors. The simulations show the influence of window dimensions and flat obstruction directly above the window opening affect the heat flux exposure to the building façade above. Table 4.9 lists the FDS simulations with varied window dimensions and obstructions. For consistency, the ventilation factor that controls the HRR inside the test compartment was kept approximately constant while adhering to the dimensional restrictions of the computational grid.

Table 4.9 Architectural feature of FDS demonstrated simulations

Simulation	HRR (kW)	W_v (m)	H_v (m)	$A_v\sqrt{H_v}$ ($m^{5/2}$)	n (4.16)	Obstruction (m)
1	600	0.9	0.9	0.77	2.00	None
2	600	0.5	1.3	0.74	0.77	None
3	600	1.3	0.7	0.76	3.71	None
4	600	0.9	0.9	0.77	2.00	0.2
5	600	0.9	0.9	0.77	2.00	0.2
6	900	0.9	0.9	0.77	2.00	None
7	900	0.5	1.3	0.74	0.77	None
8	900	1.3	0.7	0.76	3.71	None
9	900	0.9	0.9	0.77	2.00	0.2
10	900	0.9	0.9	0.77	2.00	0.2

To accurately represent the fire dynamics created by the façade window opening, Zhao (2016) suggested a series of non-dimensional length scales that may be used to correctly size the cells of the computational grid in an FDS simulation. These length scales are ratios between the simulation grid size and the non-dimensional fire diameter, a façade window length scale and the window hydraulic diameter. The three ratios are given as

$$\frac{\overset{*}{D}}{dx} = \frac{\left(\frac{Q}{\rho_\infty T_\infty C_p \sqrt{g}}\right)^{2/5}}{dx}, \tag{4.33}$$

$$\overset{*}{l_1} = \frac{l_1}{dx} = \frac{\left(A_v\sqrt{H_v}\right)^{2/5}}{dx}, \tag{4.34}$$

$$\overset{*}{l_2} = \frac{D_h}{dx} = \frac{\frac{2W_vH_v}{W_v+H_v}}{dx}. \tag{4.35}$$

It is recommended that the values of these ratios are at least 10. For this demonstration, a grid size of 5 cm was selected. Figures 4.27 and 4.28 show a set of images from simulations showing the effects of window dimensions and obstructions on the shape and trajectory of the fire plume, respectively. Further, Figs. 4.29 and 4.30 show the variation of heat flux for different configurations of obstructions and window sizes, respectively.

The presence of a sufficiently large horizontal obstruction directly above the window opening reduced heat flux onto the façade surface. A small obstruction resulted in a small increase in heat flux to the façade wall. This is likely due to circulation of the plume back toward the façade and requires experimental verification. Of the three window geometries, the geometry with the largest n (4.16) generated the largest incident heat flux on the façade. The results are in qualitative agreement with the work of Peng (2016) and Oleszkiewicz (1989).

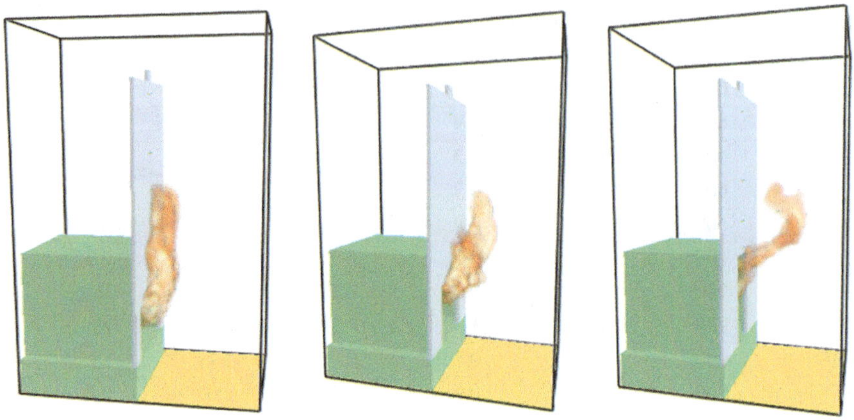

(a) $n = 0.77$ (b) $n = 2.00$ (c) $n = 3.71$

Fig. 4.27 JIS A1310 FDS simulations with different values of n

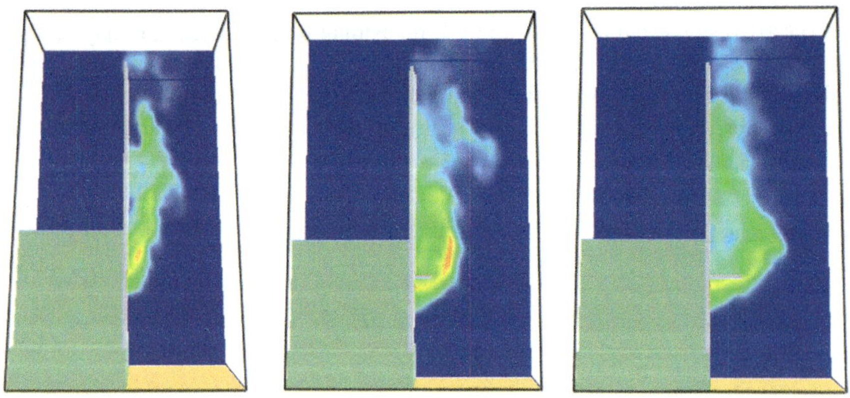

(a) No obstruction. (b) Obstruction of 0.2 m. (c) Obstruction of 0.4 m.

Fig. 4.28 JIS A1310 FDS simulations with different sizes of obstructions

A more complex example of investigating the effect of architectural designs using FDS was reported by Giraldo (2012). They studied the influence of horizontal projection and slopping window designs on the fire plume ejected from the opening as shown in Fig. 4.31. FDS results demonstrated that the horizontal projections and façade orientation can assist in pushing the fire plume trajectory away from the spandrel above the window to reduce the thermal exposure from the fire. The results from the modeling are presented in Fig. 4.32.

While they did not compare this result with any experiments, the study showed how designers may consider several options before validating with experiments.

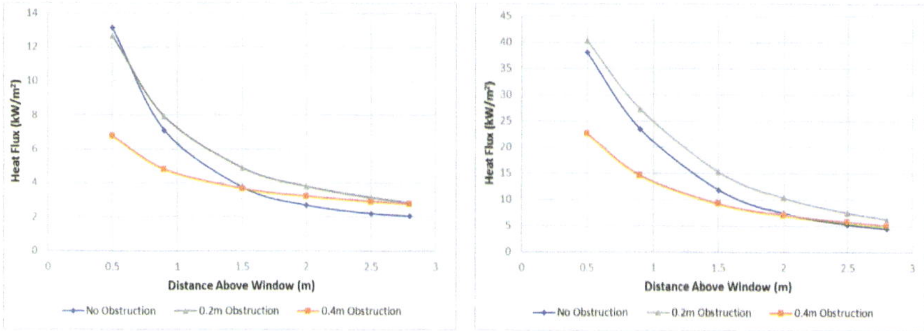

(a) 600 kW. (b) 900 kW.

Fig. 4.29 JIS A1310 FDS simulation heat flux variation with different sizes of obstructions

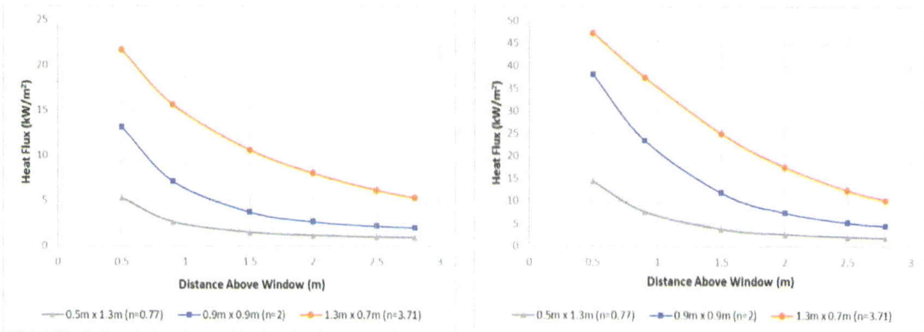

(a) 600 kW. (b) 900 kW.

Fig. 4.30 JIS A1310 FDS simulations heat flux variation with different window dimensions

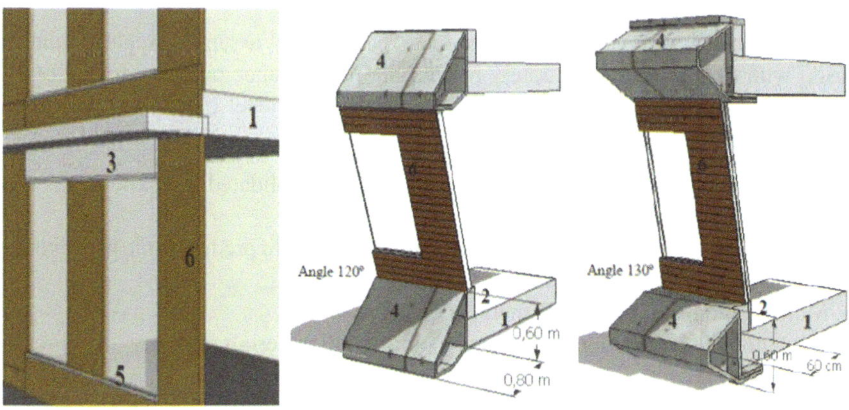

(a) Horizontal projection. (b) Sloped window 1. (c) Sloped window 2.

Fig. 4.31 FDS study of three window designs (Giraldo 2012)

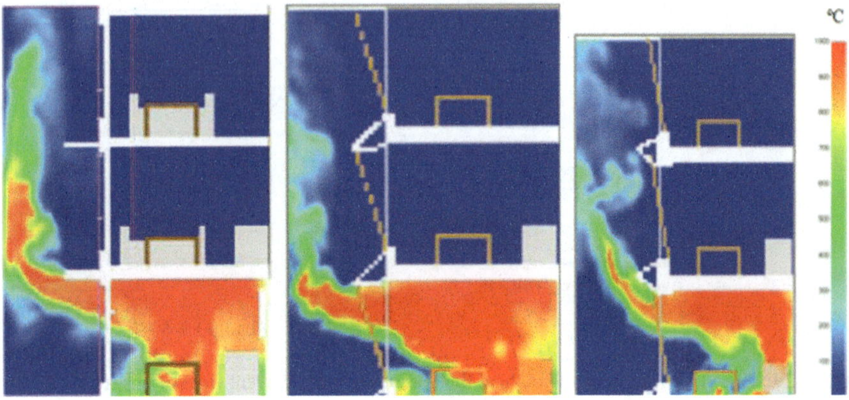

(a) Horizontal projection. (b) Sloped window 1. (c) Sloped window 2.

Fig. 4.32 Influence of window designs (Giraldo 2012)

A similar benefit of a horizontal projection was observed by Morgado (2015) who also compared their experimental results with computer simulations using FDS. However, Morgado's Fire Dynamics Simulator -based heat flux results did not correlate with the experimental work, indicating that predictive modeling of these architectural features is still in the development phase.

4.3.2 Predicting Façade System Fire Performance

Modeling of a façade system to a design fire exposure (e.g., assembly-scale regulatory tests) with the ability to accurately predict the performance outcome is one of the desired goals of computer simulations. However, this level of modeling presents several challenges. These include:

- The capability of a multi-physics model to develop validated fire exposure conditions to the façade;
- Appropriate heat transfer and combustion sub-models to predict flame propagation on materials directly exposed to fire;
- Melting and dripping of polymers; and
- Decomposition, ignition, and combustibility of materials that are not directly exposed to the fire.

Dréan (2019) provided an example of predicting ACP façade fire performance for ISO 13785-1 intermediate panel test using FDS. This study showed the sophistication required to develop the input data and identified sub-models that best describe the material response. Additionally, inferences have to be developed to estimate when the exposed aluminum surface may break open, exposing the sealant to direct flame exposure, or when and at what rate the sealant vapors (PE in this case) may escape

Fig. 4.33 Visual comparison of experimental and numerical results (Dréan 2019)

from the panel and contribute to the fire exposure. A comparison of experimental photographs and the numerical results at various times from the paper is depicted in Fig. 4.33.

It may be observed that off-gassing at the seam observed in the experiment, between 2 and 3 min, is not replicated numerically due to challenges of modeling this behavior. Further, there is evidence of pooling from the panel at 7 min, most likely from melting PE.

Blake (2018) presented a case study of comparing numerical results using FDS with experimental data from a BS 8414 test using a panel type that was installed in the Grenfell Tower. The study demonstrated the challenge of modeling a multilayer composite consisting of ACP with PE as a sealant, a void space, and thermal insulation. Part of the challenge of correlating FDS results with experiments may be the 1-D heat transfer model in FDS.

To overcome this 1-D heat transfer limitation, Wickström (2008) proposed combining the strength of FDS to define the thermal exposure on a surface and use that as input to a finite element code for heat transfer. While this increases computing burden, it overcomes some of the heat transfer limitations of the current CFD codes.

With continued improvement in computing power and the CFD codes, many of these challenges will be overcome. For regulatory acceptance, CFD modeling results will have to be validated with case studies and sensitivity analysis relative to uncertainty in input variables.

Chapter 5
Detailed Case Studies of Full-Scale Experiments

Detailed discussions of seven case studies originating from full-scale fire experiments are presented in this chapter. First, a description of the design and construction of the experimental facility is provided. Subsequently, full scale experiments are discussed with timeline of events correlated with photographs before, during, and after the experiments. Finally, a discussion on behavior and failure mechanisms of different façade and non-façade components is presented along with a mention of the main challenges faced during these experiments. In all the experiments, fire scenarios were developed using real office furniture (wood cribs, sofa sets, reception desks, curtains, carpets, etc.), and fire was not controlled and allowed to grow naturally after ignition.

5.1 Experimental Facility

This section presents details of planning, construction, fire protection, and instrumentation of the experimental facility (fire building) that was developed in 2016 at IIT Gandhinagar in collaboration with Underwriters Laboratories. While a bird's eye view of these details is available in the literature (Srivastava 2018), the discussion here is more comprehensive and is illustrated with site photographs where necessary.

5.1.1 Structural and Partition System

The main structural system of the experimental facility is a steel frame comprising of eight columns made of ISMB 600, two primary beams, and nine secondary beams made of ISMB 400, as shown in Fig. 5.1 (this figure shows only the Ground Floor (GF) view of the structural system; the same system is replicated at the upper floors).

G. Srivastava and P. D. Gandhi, *Performance of Combustible Façade Systems Used in Green Building Technologies Under Fire*, Springer Transactions in Civil and Environmental Engineering, https://doi.org/10.1007/978-981-16-3112-2_5

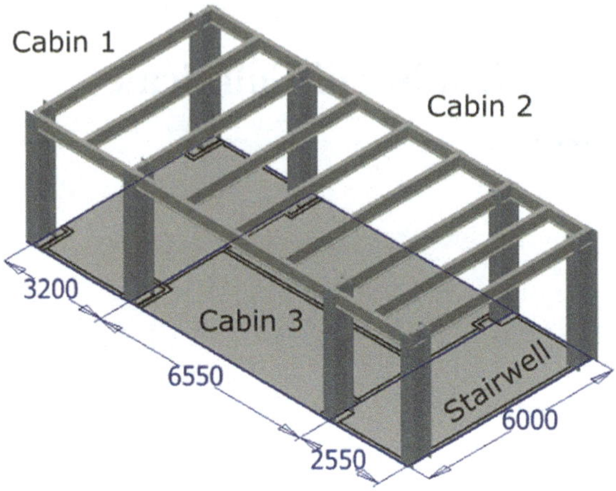

Fig. 5.1 Structural frame layout of the experimental facility. Three cabins for fire testing were created using masonry partition; one cabin was used for housing the staircase. The facility is three-storied; beam/column layout of only the GF is shown for clarity. All the dimensions shown are in mm

(a) Column base connection.

(b) Beam-column connection.

Fig. 5.2 Column base and beam-column connections

Columns are connected to the foundation through a 20 mm base plate with 10 high-strength friction grip M30 foundation bolts, as shown in Fig. 5.2a while the beam-column connections are flexible end-plate connections, as shown in Fig. 5.2b.

A deck slab is provided at all floor levels using a 1 mm thick galvanized steel deck sheet (yield strength 250 MPa, trapezoidal profile), 8 mm dia torr steel (yield strength 500 MPa) reinforcement in both directions and M25 grade concrete, as shown in Fig. 5.3.

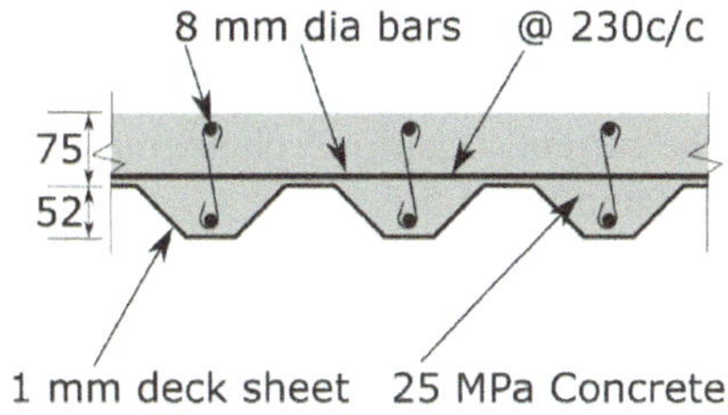

(a) Deck slab schematic.

(b) Deck sheet with reinforcements.

Fig. 5.3 Deck slab design and details

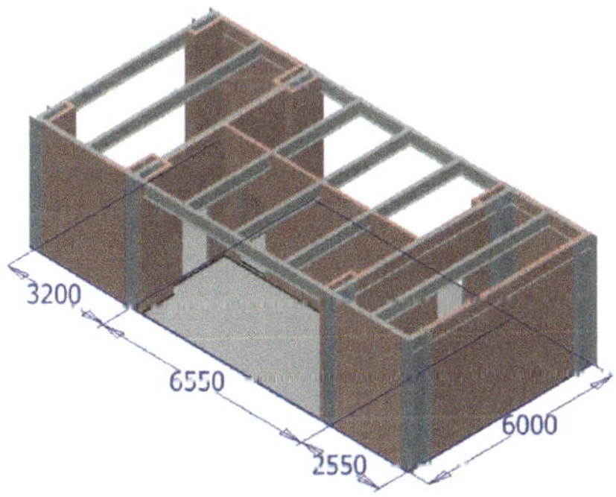

(a) Schematic of masonry partitions.

(b) Photograph of masonry enclosures.

Fig. 5.4 Use of brick masonry for internal partitioning. Faces of the building where façades are to be installed are kept open

Internal partitions are provided using 115 mm thick non-load bearing masonry walls, as shown in Fig. 5.4. 1.5 h rated fire doors are used to provide access to the cabins through the partition walls; these are 46 mm thick single swing steel stiffened doors made of 16 gauge galvanized steel incorporating single-point latches.

5.1.2 Fire Protection System

As the building utilizes a steel structural frame, fire protection is an important consideration. Fire protection of the structural columns is provided through 115 mm clay brick masonry (these masonry enclosures can be seen in Fig. 5.4a). As per NBC (2016), such a thickness of masonry wall can provide more than four hours of fire rating and hence, it was deemed suitable for the current scenario. A photograph of a partially constructed masonry enclosure is shown in Fig. 5.5.

Since the masonry enclosures are not flush with the column, the additional air gap aids in reducing the heat transfer from compartment fires to the columns.

For beams and deck slab, a two-layer board fire protection system was designed. This comprises of Plaster of Paris (POP) and cement boards. 12 mm POP boards are installed flush with the deck slab bottom, beam flanges, and webs. 18 mm cement boards are provided as a false ceiling at all floor levels, fixed to the beam bottom through 1″ mild steel box pipes, as shown in Fig. 5.6. As per NBC (2016), such cement boards are expected to provide a 30min fire rating, and the designed two-layer system was found to be suitable for the experiments. There are inherent gaps between two successive cement boards due to construction tolerances; these are filled with POP paste before the experiments. One practical aspect of the installation of the boards required consideration of the spacing of the screws used to fasten the cement boards to the MS box pipes as the size of the cement board available in the market was fairly large (8′ × 4′). An 8″ spacing of the screws was found to be suitable for the experiments (in one of the experiments, 20″ spacing was provided which caused

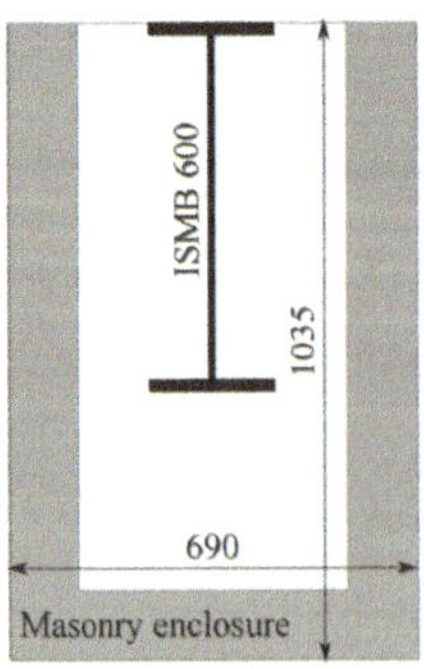

(a) Design of fire protection.

(b) Photograph of column fire protection.

Fig. 5.5 Masonry enclosures used to provide fire protection to the main structural columns of the experimental facility. Partially constructed internal partition wall can also be seen. Holes are kept in the masonry enclosures in order to facilitate wiring for instrumentation

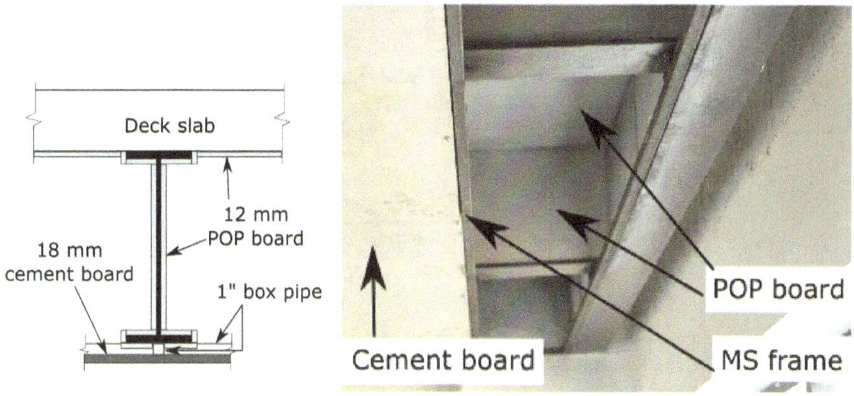

(a) Design of board protection. (b) Photograph of board protection.

Fig. 5.6 Details of the two-layer board protection provided for deck slab and beams. POP is installed flush with the deck slab and beams. Cement board is installed as a false ceiling, on a mild steel (MS) box pipe fitted along the bottom flange of beams

early failure of the cement boards and POP boards subsequently leading to direct fire exposure of deck slab and beams). The board protection system is essentially sacrificial in nature and is required to be replaced before every new experiment. The masonry protection system is more robust, and only the external plaster is required to be refurbished before a new experiment.

In addition to the passive fire protection for the structural system, the experimental facility is also provided with an external fire hydrant system of dry riser type to enable internal firefighting as and when necessary. It is expected that this provision may not be required in the planned experiments as after the façade system fails during a fire, it provides ample opening to allow firefighting operations from outside of the building.

5.1.3 Instrumentation

The following types of instrumentation have been planned for the building to enable data collection during the fire experiments.

- thermocouples (K type) to measure fire and structural temperatures;
- internal video cameras to obtain internal videos that enable correlation of temperature data and smoke ingress with events;
- Linear Variable Differential Transformer (LVDT) for measurement of out-of-plane deformations of the partition walls;
- strain gauge to measure strain levels in structural beams and columns;
- external video cameras to obtain external video footage; and

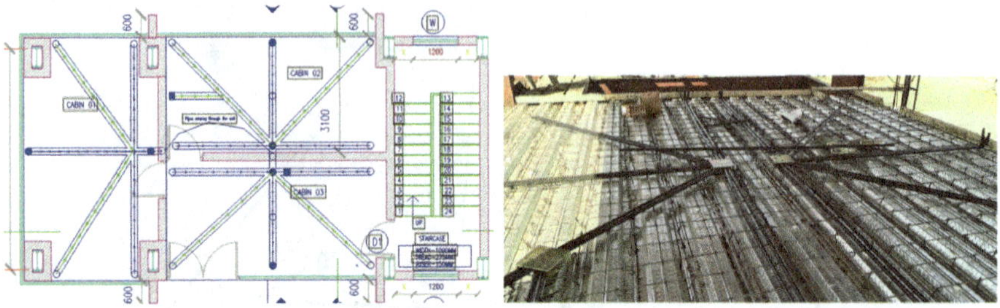

(a) Scheme of conduits for wiring. (b) Placing of the conduits for wiring.

Fig. 5.7 Scheme and placing of conduits for wiring of the sensors. Junction boxes are provided for ease of access. The conduits are continued vertically from one floor to the next through walls

- external thermal imaging camera for recording of façade temperatures from outside the building.

Installation of internal sensors (thermocouples, cameras, LVDT and strain gauges), require wiring of the sensor cables through the different cabins and hence, protection from fire during experiments. A networked system of conduits was designed for each floor to enable wiring, as shown in Fig. 5.7. For interstory wiring, conduits are taken either through the walls or through the cavity available between structural columns and masonry enclosures. While these conduits are useful for wiring of sensors placed at the floor level, wiring of sensors to be placed below the ceiling level is carried out through the space between the false ceiling of cement board and the deck slab. In certain experiments, thermocouple trees are provided in the fire compartments. While the wires could be taken up to the tree through one of the conduits mentioned earlier, the tree itself is made using a 1″ box pipe (mild steel). The outer surface of the box pipe, and the exposed wires of the sensors are coated with intumescent paint to avoid burning of the wires during fire exposure.

The overall instrumentation layout is shown in Fig. 5.8. Length of wires of sensors can pose limitations due to the fact that longer wire lengths cause greater drop in voltage and can lead to erroneous readings. The manufacturer of the K-type thermocouples (Inconel junction) specified a maximum wire length of 12 m for acceptable attenuation levels. Consequently, the data acquisition system (Data Acquisition System (DAQ)) is required to be placed inside the building. To optimize wire length of thermocouples across different floors, the DAQ is placed at the first floor (FF) and in the compartment opposite to the site being tested. The potential fire compartment (once fire reached FF from the ignition compartment) is separated from the DAQ through the brick masonry partition wall, described earlier, and a fire door. The sensor wires are routed through the floor/wall conduits or through the false ceiling for connecting to the DAQ. The wires are first moved horizontally (i.e., to the next compartment in the same floor) and then vertically (in order to reach FF where DAQ was placed); this ensures optimal lengths of the wires. Since no one could be present inside the building to operate the DAQ, it is operated remotely through a computer kept in a control room (30 m away from the fire building); the DAQ is connected

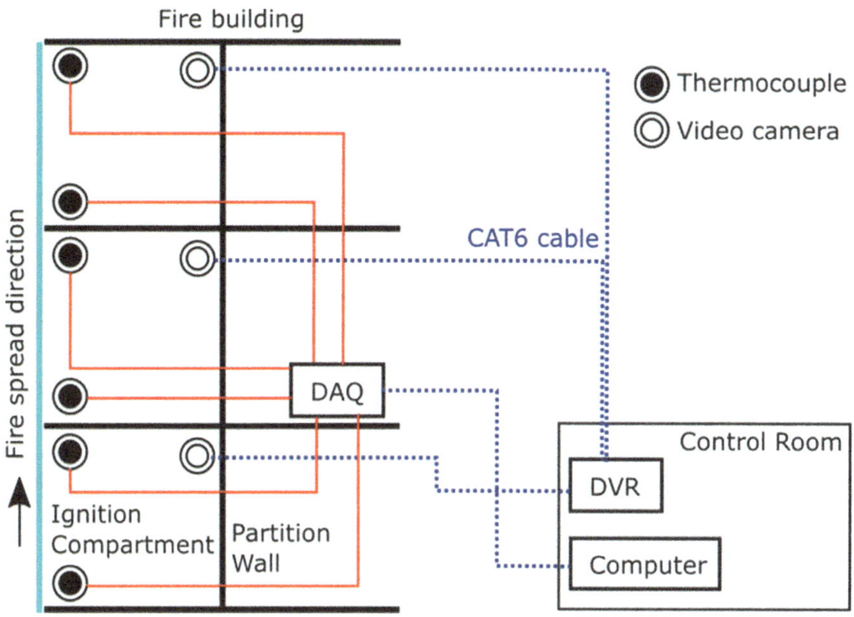

Fig. 5.8 Instrumentation layout. Thermocouples and video cameras are installed in all compartments of interest in the fire building. DAQ is kept inside the building due to constraints on the sensor wire lengths. DAQ is controlled remotely through a computer kept in the control room; digital video recorder (DVR) is also kept in the control room

to the computer through a CAT6 cable (Ethernet cable). The power supply for the DAQ is also provided from the control room. The power and CAT6 cables are routed through different conduits to avoid electromagnetic interference. LVDT and strain gauge can also connected to the same DAQ using a similar wiring approach. DAQ of 128 channels (National Instruments, PXIe 8840) is used. While the DAQ supports capturing data at the rate of 90 samples per second, in most experiments, 10 samples per second capture rate was used. Video cameras (both internal and external) are connected to a DVR kept in the control room.

Thermocouples being used are of K type with Inconel 600 sheathing having 4.5 mm diameter (inclusive of sheathing). Bead diameter is 1.2 mm with a temperature measurement range of 0–1200 °C. High-resolution video cameras with a rated operating range from −10 to 60 °C are used inside and outside the building. Internal cameras are expected to fail once the fire size became significant as they are not protected. However, initial smoke and fire movement visuals can be captured before the cameras fail or the compartment becomes full of smoke. For thermal imaging, FLIR T360 infrared camera is being used from outside the building.

5.1.4 *Fire Scenarios*

The experimental facility is designed to study two types of fire spread mechanisms, as shown in Fig. 5.9. Cabins 1 and 2 are designed to study external fire spread mechanism wherein the façade fails first and the fire spreads to the next floor externally—the so-called leap-frog effect. Cabin 3 employs masonry spandrel to enable a more fire resistant spandrel area and is designed to study internal fire spread mechanism due to the failure of firestop. It is to be noted that Cabins 1 and 2 can potentially undergo an internal fire spread mechanism in case the façade is resistant enough to fire and the firestop fails first. This is possible when using non-combustible façade panels such as glass. Cabins 1 and 2 provide greater compliance near the spandrel area and as such the effects of hot movements of the cladding frame on the firestop can be observed. Additionally, design of the façade side of each cabin allows sufficient flexibility to experiment with different kinds of façade installations. For instance, in an experiment that involved a system of insulation and ACP façade, the side of Cabin 2 was modified with additional masonry walls on the façade side to install the insulation boards.

A photograph of the finished experimental facility just before a series of experiments is shown in Fig. 5.10.

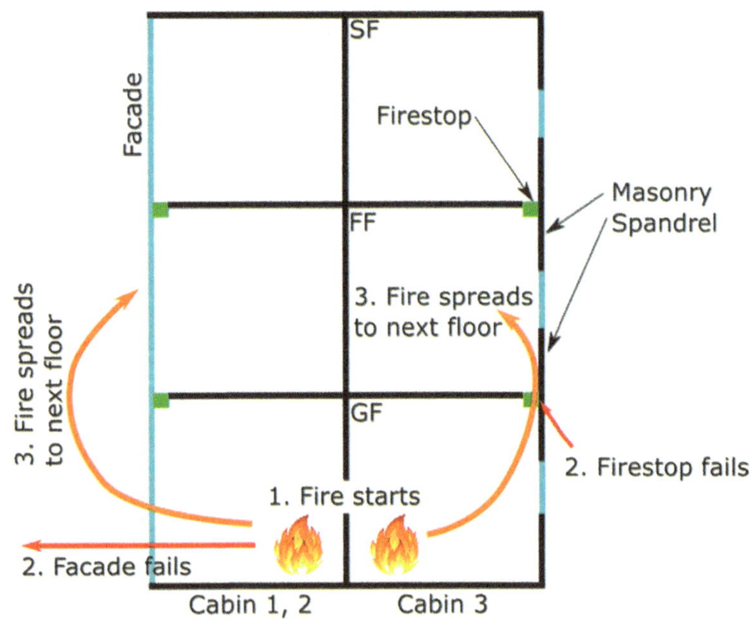

Fig. 5.9 Fire spread mechanisms. Cabins 1 and 2 designed to study external spread mechanism; Cabin 3 designed to study internal spread mechanism. The sequence of events shows the two types of fire spread mechanisms

(a) During façade installation. (b) Finished view.

Fig. 5.10 Photographs of the experimental facility. The left façade face is Cabin 1, used to study a glass+ACP system. The right side is Cabin 3 with masonry spandrel, used to study internal fire spread mechanism

5.1.5 Façade Installation

In general, façade is installed using the conventional stick system wherein aluminum cladding frame is first fixed to the building through MS brackets and subsequently, façade panels (glass, ACP, MDF) are fixed to the cladding frame through pressure plates/mechanical fasteners. Rectangular aluminum tubes of size 5864 mm are used as the cladding frame (mullions and transoms). Cladding frame is secured at all floor levels to the deck slab through L-shaped 10 mm thick MS brackets and anchor fasteners. Façade panels were usually secured to the cladding frame by means of pressure tapes (with acrylic-based adhesive having melting point of 160 °C and average shear stress capacity of 0.6 MPa). Structural silicon sealant consisting of inorganic siloxane polymer and filler, cross-linker and plasticizer was used, which provides good strength up to a temperature of about 200 °C beyond which the polymer breaks down. In certain experiments, glass panels of the façade were secured through mechanical fasteners. A typical detail of the façade installation is shown in Fig. 5.11. The gap between deck slab and the façade (typically in the range of 100–250 mm) is filled with a suitable firestop system designed to prevent passage of smoke and hot gases. Figure 5.11b shows typical details of the firestop system. It comprises of compressed mineral wool in the safing area (gap between deck slab and façade) supported on Z-clips fastened to the deck slab. Depending on the compression density of the mineral wool, the firestop becomes less or more porous to the hot gases. The spandrel panel below and above the floor level is provided with fire insulation using mineral wool, as shown, and in some cases, a stainless steel backpan is used to secure the spandrel area mineral wool with appropriate density. It is important to provide insulation to not just the façade panels but also to the cladding frame in the spandrel area. As per prevailing practice, spandrel fire insulation is usually provided from inside of the

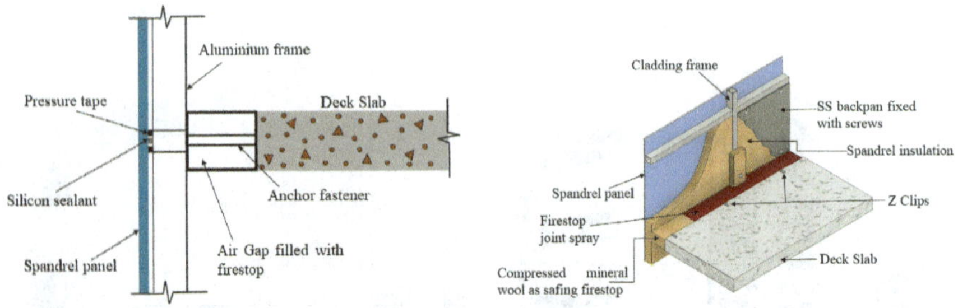

(a) Schematic of façade installation. (b) 3D view of façade installation.

Fig. 5.11 Typical installation details of façade

compartments and it is advised to use non-combustible panels in the spandrel area; however, as will be demonstrated later, consideration of fire protection of outside of the spandrel area will likely enhance the fire safety of the building.

5.2 Full-Scale Experiments

This section outlines the key features of the seven case studies (full-scale experiments) presented in this chapter. The first two experiments (Case 1 and Case 2) utilized the same façade and firestop configurations comprising of glass-ACP; the only difference among these was that in Case 1, ignition was done outside the building near GF while in Case 2, ignition was done inside the GF compartment. Both these experiments were carried out on 'Cabin 1' side of the fire building. Case 3 is similar to Case 2 in terms of ignition and other characteristics except for two differences: first, MDF was used in façade instead of ACP in combination with glass and second, it was performed on 'Cabin 2' side of the fire building.

Case 4 and Case 5 are both internal spread mechanism experiments and were performed on 'Cabin 3' side of the fire building. The only difference in these experiments was the difference in the firestop system. In Case 4, unrated firestop comprising of ply wood was used. In Case 5, rated firestop system comprising of compressed mineral wool and smoke sealing joint spray was used.

Case 6 and Case 7 involved a ventilated façade system with PIR boards as the insulation over the wall surface, an air cavity, and a cladding of glass and ACP. There were two main differences between these two experiments. The major difference was the use of a façade sprinkler system in Case 7 to assess its effectiveness. The other difference was that Case 6 experiment was performed on 'Cabin 2' side of the building while Case 7 experiment was performed on 'Cabin 1' side of the fire building. A summary of all the experiments is provided in Table 5.1.

Table 5.1 Summary of all full-scale experiments

Case	Façade	Firestop	Spandrel	Ignition
1	Glass-ACP	Rated	Combustible	External
2	Glass-ACP	Rated	Combustible	Internal
3	Glass-MDF	Rated	Combustible	Internal
4	Glass with masonry spandrel	Unrated	Non-combustible	Internal
5	Glass with masonry spandrel	Rated	Combustible	Internal
6	Glass-ACP with PIR insulation and air cavity	Rated	Combustible	Internal
7	Glass-ACP with PIR insulation and air cavity; façade sprinklers were also installed	Rated	Combustible	External

5.3 Case 1: Glass-ACP System—External Ignition

A 60% glass and 40% ACP proportion was utilized to make the façade assembly, as shown in Fig. 5.12a. A wood crib (made of 256 pinewood members of square cross-sectional dimensions of 38 mm and 1.22 m length arranged in 16 layers of 16 members each) placed outside the building, as shown in Fig. 5.39b, was used as the ignition source to simulate an external fire scenario for the façade. Two panes of the façade at GF were opened to provide initial ventilation, corresponding to an initial ventilation factor of $F_v = 0.01\ m^{0.5}$. No initial ventilation was provided at FF and SF. As the fire progressed and the façade panels failed, ventilation factor increased.

Fire scenarios of this experiment were developed using realistic furnishings, as shown in Fig. 5.13 with a quantitative summary of the fire loads provided in Table 5.3. To initiate the fire, gasoline with total energy content of 21MJ was placed in a steel pan next to the external wood crib and ignited. The fire was then allowed to grow naturally until it reached the SF. A detailed account of the progress of fire with major events is discussed next (Table 5.2).

5.3.1 Behavior of Fire and Façade with Timeline of Events

5.3.1.1 Event 1: Pre-Checks (−5 min)

Some pre-checks were performed on the systems to ensure their functionality. A few randomly chosen thermocouples were exposed to small flames (through a lighter)

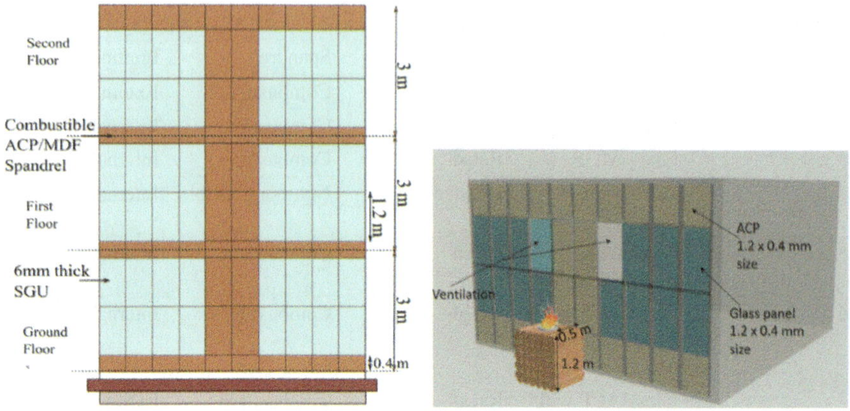

(a) Façade composition. (b) Ignition and ventilation setup.

Fig. 5.12 Façade composition, ignition, and initial ventilation setup for Case 1

(a) Ground Floor (b) First Floor (c) Second Floor

Fig. 5.13 Fire loads used in Case 1

to qualitatively assess that the corresponding channels showed corresponding rise in temperature. The DAQ was initiated five minutes before ignition (which is the reason that the temperature-time curves in Fig. 5.22 show ambient temperature levels in the first five minutes.

5.3.1.2 Event 2: Ignition (0 min)

The wood crib placed outside the façade at the GF was ignited through a flame torch. Since the wood crib was placed outside the compartment, its individual fire behavior was expected to remain a fuel-controlled one before it would begin interacting with the combustible façade components. Initially, the fire was small and no temperature rise was observed inside the compartments. Snapshots of the fire building before and after the ignition are shown in Fig. 5.14.

Table 5.2 Quantitative summary of fire loads of Case 1

S. No.	Item	Material	Quantity	Fire load (MJ)
Ground floor				
1	Chair	Wood, polyurethane	3	686.4
2	Table	Wood	1	440.0
3	Mattress	Cotton	3	1042.8
4	Curtains/sheets	Cotton	–	126.4
5	Crib	Pine wood	1	1722.2
6	Bench	Wood	1	580.8
	Total			4598.5
First floor				
1	Chair	Wood	1	96.8
2	Table	Wood, steel	1	633.6
3	Mattress	Cotton	2	695.2
4	Curtains/sheets	Cotton	1	79.0
5	Crib	Pine wood	1	1722.2
6	Cupboard	Wood	1	440.0
	Total			3634.2
Second floor				
1	Curtains/sheets	Cotton	1	79.0
2	Cupboard	Wood	1	704.0
	Total			783.0

5.3.1.3 Event 3 (2.2 min)

The fire of the wood crib was fully developed, and the flames began to interact with the façade and the GF compartment. The flames reached inside the GF compartment through the initial openings that were provided. Consequently, the thermocouples installed at the roof-level of the GF showed rise in temperature. However, none of the items of the compartment caught fire at this stage; the remaining thermocouples showed ambient temperature readings.

5.3.1.4 Event 4 (2.75 min)

Shortly after the flames began reaching inside the GF compartment, the curtains (installed next to the façade) caught fire. The roof-level temperatures near the façade rose to around 200 °C while the far end thermocouple registered around 100°C temperature. A reasonable amount of smoke also penetrated (and some generated due to combustion of the curtains) inside the GF compartment. At the FF and SF levels, the temperatures were still at ambient levels while mild smoke obscuration was observed.

(a) Before ignition. (b) Just after ignition.

Fig. 5.14 Photographs of the building just before and after ignition

Views from the outside and inside of the building are shown in Fig. 5.15. The smoke obscuration at FF and SF levels indicated the presence of small gaps in the façade system (either at the building-façade joints or at the panel-façade frame joints). This could also have been due to the onset of hot movements and the inability of the firestop system to cope with it (since the spandrel area was made of combustible panels).

5.3.1.5 Event 5 (3.03 min)

Fire intensity at the GF increased to an extent that the internal lights were burnt. Smoke intensity had also enhanced significantly in the last 15 s. The average temperatures at GF level were about 200 °C while temperatures at the FF and SF levels were still at ambient levels. Smoke obscuration was observed at SF also indicating localized failure of the FF/SF firestop due to hot movements or partial failure of the façade joints. Figure 5.16 shows that external flames were reaching up to the SF floor level. Several luminous streaks were observed which were due to the dripping polymeric core of the ACP. The internal compartment fire of GF was also visible from outside.

(a) View from outside.

(b) View inside GF.

(c) View inside FF.

(d) View inside SF.

Fig. 5.15 Views of different portions of the building at event 4

5.3.1.6 Event 6 (3.13 min)

Lights at the FF and SF went off during this time although the temperatures at both these floors remained at ambient levels; thus, the lights went off primarily due to burning of the power cable that extended from GF to the upper floors for the light fixtures. At the GF, the fire expanded from one side of the façade to the other side, horizontally, primarily through the curtains. The maximum temperature near the façade at GF rose to 400 °C.

(a) View from outside. (b) View inside GF.

(c) View inside FF. (d) View inside SF.

Fig. 5.16 Views of different portions of the building at event 5

5.3.1.7 Event 7 (3.33 min)

The entire curtain of the GF was aflame, and consequently, the maximum temperature at the roof-level was about 550 °C. It is to be noted that roof-level temperature implies temperature below the false-ceiling at the roof level. As described earlier, a two-layer fire protection was installed for the steel beams and deck slab: first, a layer of POP boards which were snug to the beam web/flange and deck slab and second, a false ceiling made of cement boards. The temperature at the beam level (above the cement board false ceiling) at this stage was about 19 °C, i.e., at ambient levels. This indicated the efficacy of the fire protection system. At the FF, the roof-level temperature (below false-ceiling) rose to about 33 °C while the beam-level temperature (above false ceiling) was 19 °C. The rise in the temperature at the FF level could be attributed to the passage of hot gases from GF to FF. The SF did not show any significant temperature rise up to this point.

5.3.1.8 Event 8 (3.5 min)

Curtains at the GF level were fully engulfed in flames and reasonable amount of hot gases had collected at the roof level at GF. Due to the feedback from the roof-level hot gases, temperature at the base level of GF increased to 40 °C. The overall average temperatures at all the floors did not change much since the last event.

5.3.1.9 Event 9 (3.67 min)

The wood crib kept inside the GF caught fire at this stage, indicating a significant radiant heat feedback from the flames and the hot upper layer of gases. The GF was completely filled with smoke. The glass panel below the ventilation opening fell down due to excessive distortion of the surrounding aluminum frame and failure of the sealing material (at the glass–frame interface). Roof-level maximum temperature at GF was about 675 °C while that at the base level was about 50 °C.

5.3.1.10 Event 10 (4.58 min)

The entire GF was under fire at this stage with maximum roof-level temperature at 685 °C. The temperature at the beam-level (above false ceiling) was 25 °C, indicating the effectiveness of the fire protection system. The flames of the external wood crib along with the hot gases from GF caused propagation of the fire at the FF. The roof-level temperature near the façade at the FF was about 150 °C. Temperatures at the SF were still at the ambient level although smoke obscuration increased significantly. Views from outside and inside the FF are shown in Fig. 5.17. Enhanced size of the outside fire can also be observed due to the continuous addition of fire load due to dripping of the polymeric core of ACP.

5.3.1.11 Event 11 (4.95 min)

Three more glass panels fell (in intact form) at the GF level due to failure of the binder sealant and expansion of the cladding frame, which enhanced the ventilation conditions causing greater ingress of fresh air. This increased the roof-level temperatures near the façade to about 740 °C, which were greater than the melting point of aluminum (of which the cladding frame was made). The maximum temperature at FF level rose to 190 °C.

(a) View from outside. (b) View inside FF.

Fig. 5.17 Views of different portions of the building at event 10

5.3.1.12 Event 12 (5.45 min)

At the GF, maximum temperature at the roof and the base level was about 760 and 275 °C, respectively. Temperature at the beam level was 60 °C. Curtains at the FF caught fire as fire penetrated inside the FF; this also caused the onset of failure of the façade panels at the FF thereby increasing the ventilation factor at the FF. The roof-level temperature at FF reached 280 °C. The SF was still at ambient levels with maximum roof-level temperature around 38 °C.

5.3.1.13 Event 13 (6.5 min)

Roof-level temperature at GF reduced to 670 °C due to the fire becoming ventilation-controlled. ACP sheets at FF caught fire due to an extended exposure of the ACP sheets to high temperatures which initiated fire in the inner polymeric core of ACPs. The dripping of this polymeric material caused the onset of secondary fires outside the building at the GF. Maximum temperature at the FF was about 340 °C (temperature at the beam-level was about 20 °C). None of the items within the FF caught fire at this stage; temperature at the SF reached about 48 °C primarily due to passage of hot gases. Figure 5.18 shows views from outside the building and inside the FF.

(a) View from outside. (b) View inside FF.

Fig. 5.18 Views of different portions of the building at event 13

5.3.1.14 Event 14 (10.5 min)

GF base-level temperatures remained constant since last three minutes while the maximum roof-level temperature reached 816 °C. The temperature at the beam-level reached 120 °C indicating localized failures of the cement board false ceiling. Flashover occurred at the GF level around this time as a result of the enhanced ventilation factor due to the continuous failure of the façade system. The cladding frame melted completely on the side where the wood crib was placed. Temperatures at the FF were constant at about 300–350 °C. All the ACP sheets and glass panels at the FF failed due to excessive frame distortion and combustion of the ACP core material. The wood crib placed at the FF caught fire while the other items of FF did not catch fire yet. This can be seen as two independent fires visible in Fig. 5.19. Temperature at the beam level of FF reached 36 °C indicating no damage to the false ceiling of the FF. The roof-level temperature at the SF reached 54 °C although fire did not reach the SF yet.

5.3.1.15 Event 15 (11.62 min)

Temperature at the GF began to reduce as the fire became fuel-controlled with the fuel quantity reducing. Maximum temperature at the roof level was 710 °C while that near the beam was about 150 °C at the GF. Local failures in the cement board false ceiling were also observed at the FF. Further, the wood table kept besides the crib in

Fig. 5.19 Outside view of the building at event 14

the FF also caught fire at this stage, taking the compartment temperature at the FF to about 400 °C. No significant changes were observed at the SF

5.3.1.16 Event 16 (16.67 min)

The fire at GF diminished by itself due to the fuel being fully consumed with the maximum temperature at 305 °C. The maximum temperature at the beam-level reached 226 °C before it began to reduce (this is a well-known phenomenon of the structural temperatures lagging behind the fire temperatures). Flashover occurred at the FF during this time interval with maximum temperature reaching around 675 °C, and the compartment temperatures went beyond the melting point of aluminum; a certain portion of the cladding frame dislodged from the FF. Major failure of the cement board was also observed at the FF with the beam-level temperatures reaching as high as 510 °C at this stage. Fire did not propagate to the SF, and the maximum temperature at SF remained around 56 °C. View from the outside at this stage is shown in Fig. 5.20.

Fig. 5.20 Outside view of the building at event 16

5.3.1.17 Event 17 (18.55 min)

Fire propagated to the SF through curtains and the temperatures reached 226 °C near the façade and 140 °C at the far end. Temperature at the FF began to drop due to limitation of the availability of the fuel. A sustained secondary fire outside the GF compartment can also be seen in Fig. 5.21.

5.3.1.18 Event 18 (21.83 min)

Firefighting operations began at this stage as the stopping criterion of the experiment protocol (fire reaching the SF) was reached.

5.3.2 Time-Temperature Data

The maximum temperatures recorded at all floor levels during the experiment are shown in Fig. 5.22. No significant temperature rise was observed up to about 7 min as the fire source was outside the compartment (and ignition happened at the 5 min mark). Subsequently, a steep temperature rise was observed as the external fire (of the wood crib) was fully developed by that time. Further, the cumulative ventilation

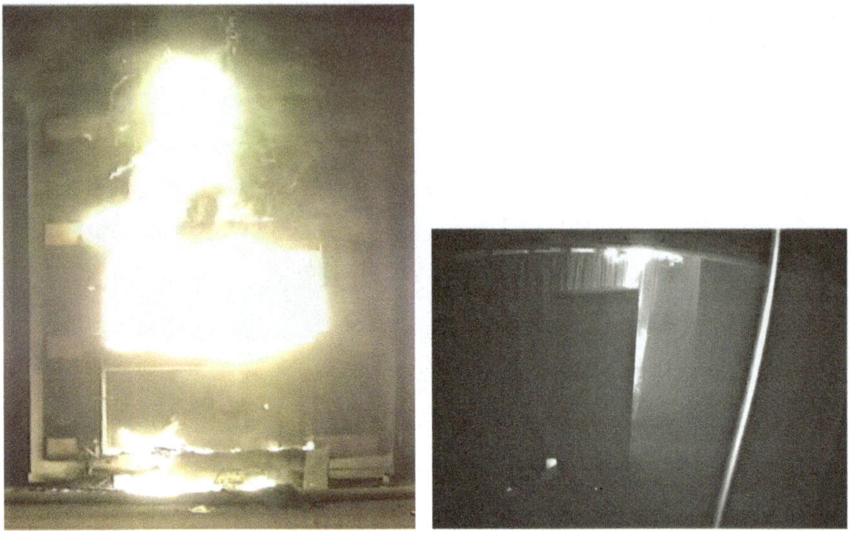

(a) View from outside. (b) View inside SF.

Fig. 5.21 Views of different portions of the building at event 17

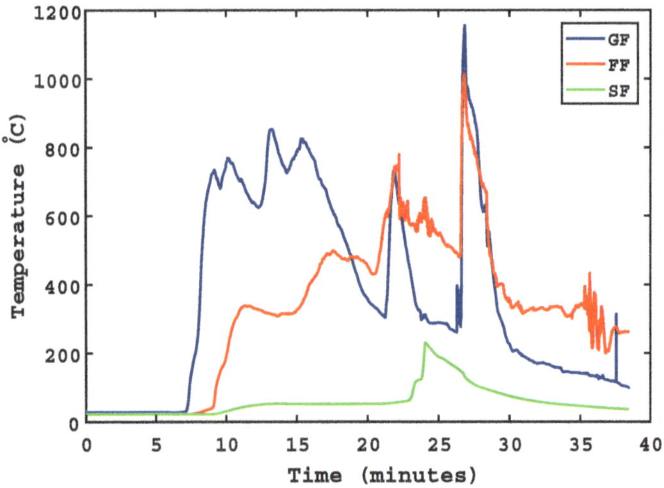

Fig. 5.22 Maximum temperatures observed at different floor levels in Case 1

factor, F_v, is shown in Fig. 5.23 which indicates that within 5 min, the entire façade of GF failed while within 10 min, façade of FF also failed. Façade of SF failed only to a limited extent as the firefighting operations were initiated.

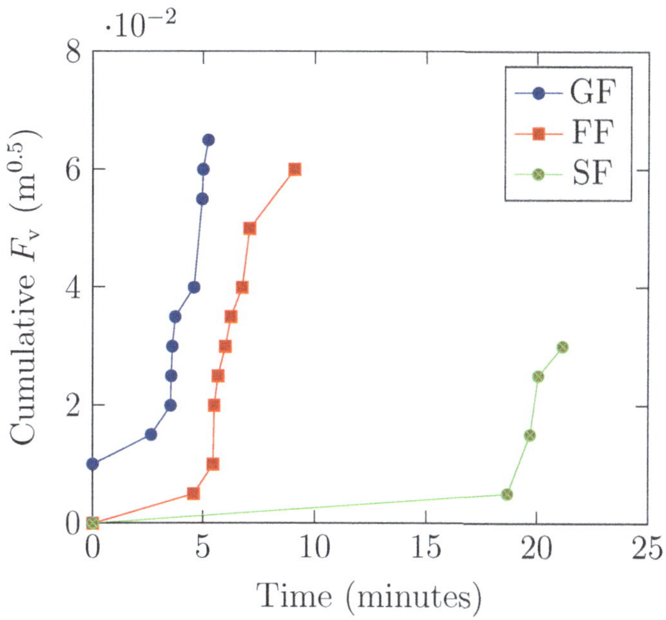

Fig. 5.23 Opening factor F_v versus time for Case 1

5.4 Case 2: Glass-ACP System—Internal Ignition

This experiment was almost identical to Case 1 except that the ignition source was inside GF (instead of being outside). The façade design and composition were identical. Figure 5.24 shows the fire load placement for this experiment, while a quantitative summary of the fire loads is given in Table 5.3.

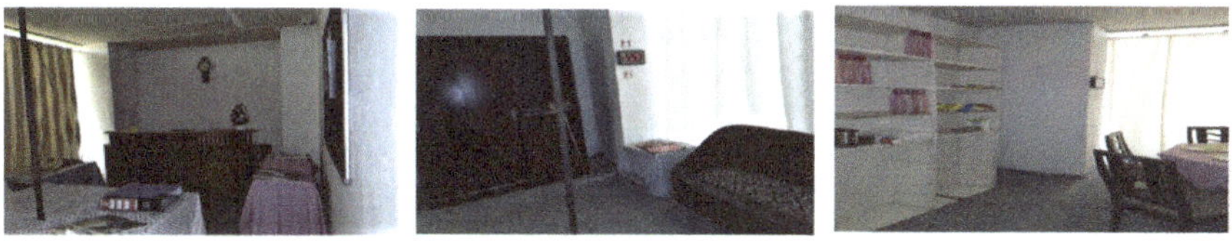

(a) Ground Floor (b) First Floor (c) Second Floor

Fig. 5.24 Fire loads used in Case 2

Table 5.3 Quantitative summary of fire loads of Case 2

S. No.	Item	Material	Quantity	Fire load (MJ)
Ground floor				
1	Chair	Wood, polyurethane	2	457.6
2	Table	Wood	1	440.0
3	Reception desk	Wood	1	1320.0
4	Curtains/sheets	Cotton/polyester	–	59.8
5	Crib	Pine wood	1	1720.9
6	Bench	Wood	1	264.0
7	Sofa	Wood, polypropylene	2	9020.0
8	Newspaper, magazines	Cellulose	–	53.6
9	Cushions	Polyester	–	715.0
	Total			14020.9
First floor				
1	Chair	Wood, polyurethane	1	228.8
2	Table	Wood, steel	1	633.6
3	Cupboards	Wood	3	2728.0
4	Curtains/sheets	Cotton	–	51.9
5	Crib	Pine wood	1	1720.9
6	Sofa	Wood, polypropylene	1	2288.0
7	Newspaper, magazines	Cellulose	–	33.0
	Total			7684.2
Second floor				
1	Chair	Wood, polyurethane	4	915.2
2	Table	Wood	1	528.0
3	Cupboard	Wood	1	704.0
4	Curtains/sheets	Cotton	–	51.9
5	Crib	Pine wood	1	1720.9
6	Appliances	Polypropylene	3	135.2
	Total			4055.2

5.4.1 Behavior of Fire and Façade with Timeline of Events

5.4.1.1 Event 1 (−8.87 min)

The data recording was initiated and the pre-checks were performed.

5.4.1.2 Event 2 (−7.65 min)

Ignition was attempted through a remote-controlled electrical switch, which failed to ignite the fuel placed in the basket on the sofa. Temperatures were still at ambient levels.

5.4.1.3 Event 3 Ignition (0 min)

Waste papers were placed in the basket along with some gasoline and the fire was ignited manually.

5.4.1.4 Event 4 (0.63 min)

Fire transmitted from sofa to curtains. Polyester curtains were not easily catching fire. Maximum temperature observed at the GF façade face was about 218 °C and the average façade temperature was about 29 °C. Façade temperature at FF and SF were at ambient levels and the compartments were clear of smoke. Views from outside the building and inside the GF are shown in Fig. 5.25.

5.4.1.5 Event 5 (2.63 min)

GF video camera stopped functioning due to heat exposure. Smoke obscuration was observed at the FF indicating passage of combustion gases from the GF to FF, whereas the SF was clear. Flames within the compartment began to reach the ceiling level at GF. As can be seen from Fig. 5.26, ACP sheets began to shrink due to exposure to heat. Maximum temperature observed at the GF façade was about 275 °C whereas at FF and SF it was around 45.8 and 28.5 °C, respectively.

5.4.1.6 Event 6 (5.63 min)

Heavy smoke obscuration was observed at the FF and camera at the FF stopped functioning. The maximum façade temperature at GF was around 250 °C. Temperatures had become constant at around 250 °C for an interval of about 10 min as the fire was

(a) View from outside. (b) View inside GF.

Fig. 5.25 Views of different portions of the building at event 2

(a) View from outside. (b) View inside FF.

Fig. 5.26 Views of different portions of the building at event 5

in the smoldering stage within the compartment due to limited ventilation through the initially provided vent openings. Wood crib placed near the ignition source did not ignite yet. The maximum roof-level temperature was about 250 °C while that at the base-level was around 50 °C at the GF. The temperature at FF and SF was around 30 °C.

5.4.1.7 Event 7 (16.46 min)

Shrinking of the ACP sheets due to heat exposure intensified with one sheet falling-off the cladding frame, increasing the ventilation at the GF. The wood crib placed in the GF began to burn. The maximum temperature at the GF façade was about 400 °C while that at FF was about 105 °C. The ACP sheets fell due to the failure of the pressure tape used to fix it to the cladding. Post fire survey observations showed that polyethylene core laminated and sandwiched between two fine aluminum sheets was exposed to fire and charred.

5.4.1.8 Event 8 (24.5 min)

Another ACP sheet adjacent to the earlier fallen one failed, further enhancing ventilation within the GF. The maximum temperature at the GF façade was about 610 °C while that at the FF façade was about 115 °C. These façade temperatures were beyond the melting point of aluminum and caused failure of the cladding frame. Visuals of these events can be seen in Fig. 5.27.

(a) Event 6. (b) Event 7. (c) Event 8.

Fig. 5.27 Views of different portions of the building at events 6, 7, 8

5.4.1.9 Event 9 (24.63 min)

Major portion of fire protection provided to the structural beams and slab failed at the GF exposing the beams and the deck slab to direct heat. Temperatures observed were around 600 °C at the GF. It was ascertained later that the early failure of the cement board false ceiling was due to the improper fixing of the board to the ceiling frame.

5.4.1.10 Event 10 (25.63 min)

Temperatures reached at around 685 °C at GF, whereas at FF, it was around 165 °C. Maximum façade temperature at the SF was around 90 °C. Flames began to leap out from the ventilation opening and fallen ACP sheets at the GF.

5.4.1.11 Event 11 (26.88 min)

Firestop material provided in the gap between cladding framework and deck slab at the FF started to fail. POP boards near façade level failed. ACP sheets provided at the spandrel area between GF and FF started to fall-off after distortion and shrinking. Temperatures at GF reached around 770 °C while at the FF reached around 220 °C.

5.4.1.12 Event 12 (27.08 min)

Four glass panels adjacent to the earlier fallen ACP sheets fell in intact form due to the failure of the binder used to fix the glass to the cladding. Flashover occurred at the GF during this period with the maximum temperature reaching 800 °C; this further intensified the melting of the aluminum cladding at certain exposed locations and distortion of the frame at other locations.

5.4.1.13 Event 13 (28.1 min)

Flames began to erupt out of the GF compartment in a significant way due to excessive failure of the glass and ACP sheets of the façade. The dripping polymeric core of ACPs initiated a secondary fire outside the building at GF. The measured temperatures at GF were around 910 °C while that at FF reached 400 °C.

5.4.1.14 Event 14 (30.6 min)

Fire transmitted to FF façade level through the leap-frog effect and the ACP sheets caught fire. Fire did not penetrate inside the compartment and remained at the façade

(a) Event 12. (b) Event 13. (c) Event 14.

Fig. 5.28 Views of different portions of the building at events 12, 13, 14

face. Subsequently, polyester curtains at the FF caught fire but the remaining items within the FF did not catch fire. Two independent fires, one at GF (including expanded secondary fire outside the building), and one at the façade of FF upward, can be observed in Fig. 5.28.

5.4.1.15 Event 15 (33.13 min)

Flames began to reach the SF façade level with a few ACPs at the SF catching fire. Fire extinguishing operations were initiated.

5.4.2 Time-Temperature Data

In this experiment, the thermocouple data could not be recorded as the CAT6 cable connecting the DAQ to the computer in control room burnt due to premature failure of the board protection provided for deck slab and beams. Video feeds from the internal cameras of FF and SF were also lost on this account before the cameras themselves failed due to fire exposure. Thermal imaging helped correlate various events with the temperatures measured from outside the building. While this did pose a limitation of not being able to know the temperature distributions within respective compartments, it provided a very useful insight to the façade temperatures. The cumulative ventilation factor, F_v, is shown in Fig. 5.29 which indicates that internal ignition was much less severe compared to the external ignition of case 1. There was almost no failure of façade panels up to 15 min during which time, the compartment fire smoldered and remained ventilation controlled to a great extent.

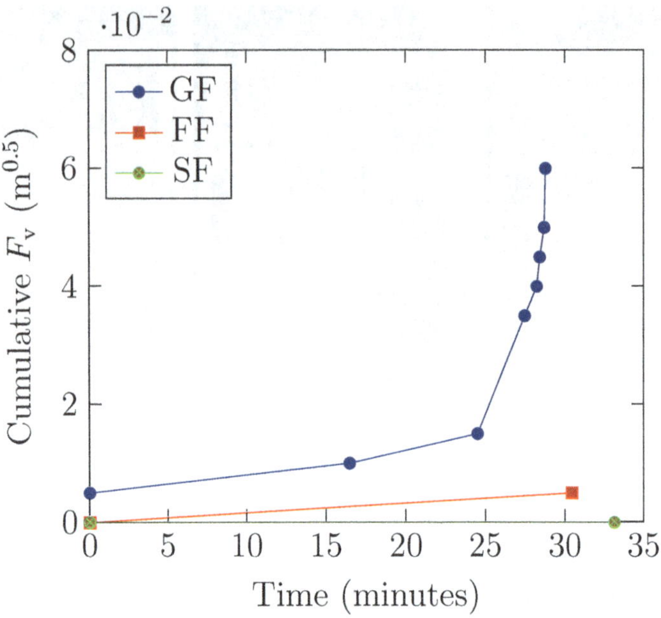

Fig. 5.29 Opening factor F_v versus time for Case 2

5.5 Case 3: Glass-MDF System

This experiment was similar to Case 2 except that instead of ACP, MDF was used in combination with glass for the façade system. Fire scenarios of this experiment are shown in Fig. 5.30 with a quantitative summary of the fire loads provided in Table 5.4.

(a) Ground Floor (b) First Floor (c) Second Floor

Fig. 5.30 Fire loads used in Case 3

Table 5.4 Quantitative summary of fire loads of Case 3

S. No.	Item	Material	Quantity	Fire load (MJ)
Ground floor				
1	Chair	Wood, polyurethane	1	228.8
2	Table	Wood	1	528.0
3	Stage furniture	wood	1	1408.0
4	Curtains/sheets	Polyester, cotton	–	55.8
5	Crib	Pine wood	2	2305.4
6	Reception desk	Wood	1	1584.0
7	Newspaper, magazines	Cellulose	1	33.0
	Total			6143.0
First floor				
1	Chair	Wood	1	228.8
2	Table	Wood	1	440.0
3	Low bench	Wood	1	880.0
4	Curtains/sheets	Polyester, cotton	–	55.8
5	Crib	Pine wood	2	2305.4
6	Cupboard	Wood	1	1408.0
7	Chest	Wood	1	528.0
8	Sofa	Wood, polyurethane	1	2288.0
9	Newspaper, magazines	Cellulose	–	33.0
	Total			8167.0
Second floor				
1	Chair	Wood	4	915.2
2	Table	Wood	1	528.0
3	Electric kettle	Polypropylene	1	48.4
4	Curtains/sheets	Polyester, cotton	–	59.8
5	Crib	Pine wood	2	2305.4
6	Plates, bowl	Polypropylene	6	87.1
7	Chest	Wood	1	528.0
	Total			4471.9

5.5.1 Behavior of Fire and Façade with Timeline of Events

5.5.1.1 Event 1: Pre-Checks (−12 min)

The initial pre-checks were performed on the systems.

5.5.1.2 Event 2 (−10.75 min)

Ignition was attempted manually in a plastic basket filled with waste paper, placed near the small wood crib. Fire source was placed in proximity to the glass and MDF assembly for easy ignition and propagation of flames.

5.5.1.3 Event 3: Ignition (0 min)

The small plastic basket fire failed to ignite the nearby wood crib and flames began to die out. A secondary ignition was done with the help of more waste papers kept near the wood crib directly. The second ignition was successful.

5.5.1.4 Event 4 (2 min)

The wood crib placed near the ignition source was ignited. The roof-level temperature above the crib reached 275 °C. Temperature at the other locations of the GF was around 30 °C. The temperature near the beam (above false ceiling) was about 23 °C, indicating proper working of the false ceiling fire protection. The maximum temperature at the GF façade level, observed through thermal imaging camera, was about 63 °C. Temperatures at the FF and SF were at ambient levels.

5.5.1.5 Event 5 (29 min)

In the 27 min duration since the previous event, the small wood crib was completely consumed by the fire but the fire did not propagate to the remaining items (e.g. large wood crib) of the GF, as can be seen in Fig. 5.31. The GF façade temperature gradually increased to 250 °C but did not cause any failure of either the cladding or the MDF panels. Due to heat and smoke, however, the internal video camera of the GF stopped working.

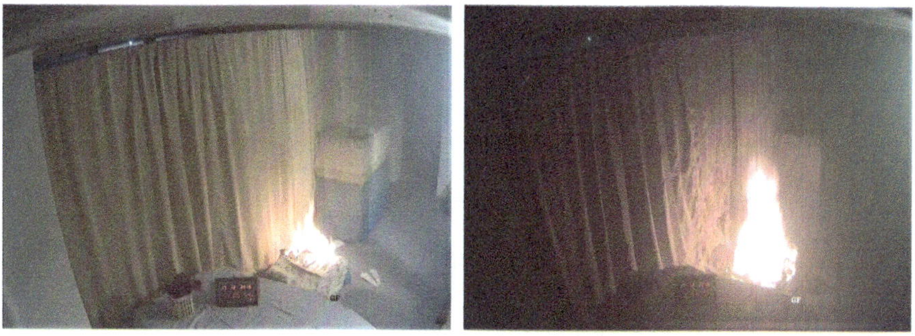

(a) At event 4. (b) Between events 4 and 5.

Fig. 5.31 View inside the GF from events 4 to 5

5.5.1.6 Event 6: Ignition (33 min)

A third ignition was carried out, this time, directly of large wood crib. After this ignition, the maximum temperature at GF was observed to be 285 °C.

5.5.1.7 Event 7 (35 min)

The large wood crib caused a more severe fire at the GF with maximum temperature reaching 475 °C. The flames reached the ceiling level. Mild smoke obscuration was observed at FF (see Fig. 5.32). No smoke obscuration happened at SF yet.

5.5.1.8 Event 8 (36.47 min)

The fire size grew along with heavy smoke production, and the flames from the wood crib erupted outside from the ventilation opening. The cement board sheet (false ceiling) directly above the wood crib fell down due to prolonged and excessive heat exposure. Temperature observed at the GF at this instant was around 630 °C. No failure of glass and MDF panels was observed.

5.5.1.9 Event 9 (41 min)

Flames from the wood crib continued to erupt out of the ventilation opening. MDF panels adjacent to the ventilation opening began to show effects of heat exposure although none of the glass or MDF panels had failed up to this time. Other fuel sources within the GF had not caught fire. Temperatures reached to about 685 °C at the GF. FF was completely obscured by smoke, while mild smoke ingress was observed at SF (Fig. 5.33).

(a) View from outside. (b) View inside FF.

Fig. 5.32 Views at event 7

(a) View from outside. (b) View inside SF.

Fig. 5.33 Views at event 9

5.5.1.10 Event 10 (44.67 min)

Flames were now penetrating through the connections of glass and MDF panels with the cladding frame, which caused failure of the sealant binder leading to falling-off of the façade panels from the cladding frame. As expected, glass panels near the ventilation opening fell first. Maximum temperature at GF reached 755 °C. It was observed that MDF panels were not catching fire (unlike ACP sheets) causing it difficult for the fire to propagate through the leap-frog mechanism. This also maintained ventilation controlled conditions for a long time within the GF which, perhaps, made it difficult for the compartment fire to develop and required multiple ignitions.

5.5.1.11 Event 11 (45.67 min)

More glass panels dislodged from the cladding frame due to the failure of the fixing mechanism. MDF panels charred heavily on one side leading to thermal bowing and subsequently, dislodged from the cladding frame with mild flames. The flames of the falling MDF panels was not sufficient to initiate or sustain a secondary fire outside the building. The removal of façade panels enhanced the ventilation conditions and invigorated the GF fire with maximum temperatures reaching 770 °C.

5.5.1.12 Event 12 (46.83 min)

Due to increased air ingress, temperatures reached around 1186 °C at the GF whereas at the GF, it was around 465 °C. Temperatures were significantly beyond the melting point of aluminum leading to melting of the cladding framework at certain locations and distortion of the frame at other locations. This led to further failure of the façade panels and more increase in the ventilation factor (Fig. 5.34).

5.5.1.13 Event 13 (48 min)

Flashover occurred at the GF with the maximum temperature reaching 1200 °C. Temperature at the FF was about 735 °C. Flames began to leap out from the GF façade in a significant manner but failed to ignite MDF panels at the FF façade and a vertical spread of fire at the façade was not observed.

5.5.1.14 Event 14 (50.5 min)

The MDF panels were now falling down with large enough flames that they initiated secondary fires at the GF outside the building. More glass panels dislodged from the cladding frame. Intensity of fire at GF began to decrease due to consumption of the fuel; the GF temperatures were around 1050 °C.

(a) Event 10. (b) Event 12. (c) Event 15.

Fig. 5.34 Views at events 10, 12, 15

5.5.1.15 Event 15 (55.75 min)

Fire load within the GF was consumed, and the fire began to die out. Fire extinguishing operations were started at the GF. Since the fire did not propagate to the FF façade, the FF façade along with the furnishings of the FF did not burn although the FF experienced temperatures in excess of 450 °C due to passage of hot gases from GF.

5.5.1.16 Event 16: Ignition (59.25 min)

Glass and MDF panels and fire loads at FF and SF were not damaged, giving an opportunity to repeat the full-scale experiment at the FF (and observe possible vertical spread to SF). Thus, the wood crib at the FF was ignited manually.

5.5.1.17 Event 17 (64 min)

The FF wood crib had ignited, and the flames erupting from the crib were hitting the ceiling. Fire also transmitted to the smaller wood crib placed adjacent to the larger one. Maximum temperature at the FF façade was around 300 °C. No effects were observed on the MDF panels.

5.5.1.18 Event 18 (66 min)

Intensity of the fire increased and heavy smoke emission was observed at the FF. The fire began to affect the fixing mechanism of the glass and MDF with the cladding

(a) Event 17. (b) Event 18. (c) Event 19.

Fig. 5.35 Views at events 17, 18, 19

frame in a manner similar to what happened during the GF fire. MDF panels began to char and shrink but did not catch fire. Maximum temperature at FF was around 550 °C (Fig. 5.35).

5.5.1.19 Event 19 (67.25 min)

MDF panel near the large wood crib caught fire. Glass panel fell in intact form due to the failure of the fixing sealant. Partial failure of the cement board above the large wood crib was observed. Maximum temperature at the FF during this time was around 850 °C.

5.5.1.20 Event 20 (69 min)

MDF panel which caught fire dislodged from the cladding. Glass panels also fell in intact form due to expansion of the aluminum cladding framework. This enhanced ventilation conditions within the compartment. Flames were observed to be leaping out of the compartment. Flashover at FF occurred during this instant. Maximum temperature at FF was around 1050 °C. Significant melting of aluminum cladding was observed.

5.5.1.21 Event 21 (70.33 min)

All MDF panels at the FF façade had burned out and all glass panels had fallen leaving the bare cladding framework. One glass panel was observed to break in-place instead

(a) Event 20. (b) Event 22.

Fig. 5.36 Views at events 20, 22

of falling out in intact form. Maximum temperature observed at the FF was around 1030 °C. All fuel had been consumed and temperature began to reduce. However, the FF fire failed to ignite the façade (and hence the internal components) of the SF (Fig. 5.36).

5.5.1.22 Event 22 (74.33 min)

The FF fire began to die due to insufficient fuel supply. The façade at the SF did not catch fire.

5.5.1.23 Event 23 (75.17 min)

Final fire extinguishing operations were initiated.

5.5.2 Time-Temperature Data

The maximum temperatures recorded at all floor levels during the experiment are shown in Fig. 5.37. The cumulative ventilation factor, F_v, for the GF ignition and fire development is shown in Fig. 5.38 which indicates that the use of MDF significantly improved the façade performance by inhibiting fire propagation and failure.

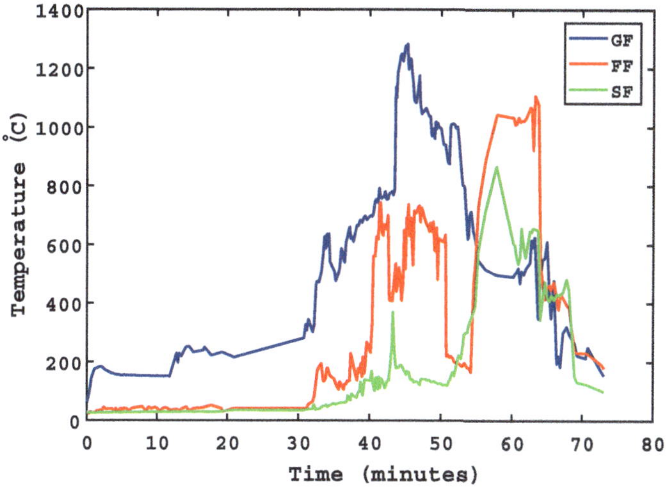

Fig. 5.37 Maximum temperatures observed at different floor levels in Case 3

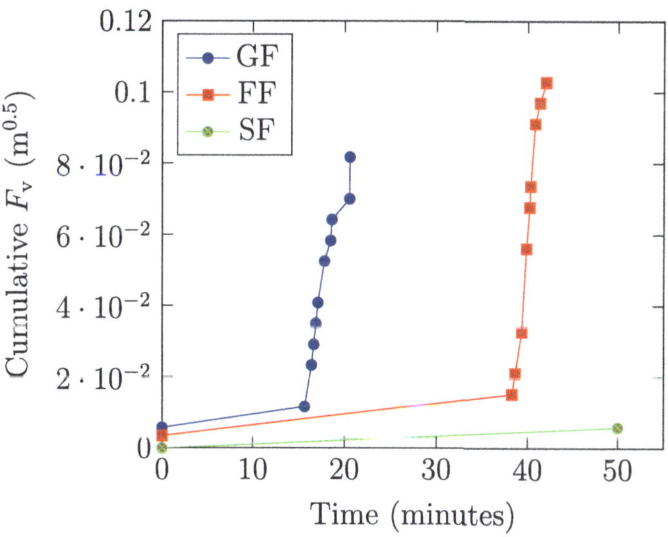

Fig. 5.38 Opening factor F_v versus time for Case 3

5.6 Case 4: Glass System with Unrated Firestop

This experiment was performed in Cabin 3 of the experimental facility (the face with masonry spandrel) to study the internal fire growth mechanism. Fire loads on all floors were kept close to the safing area. At GF, initial ventilation was provided by opening the front door ($F_v = 0.04\ m^{0.5}$), whereas no initial ventilation was provided

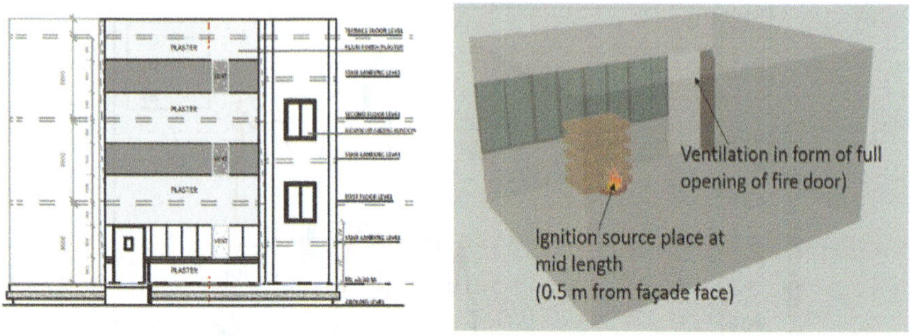

(a) Façade composition. (b) Ignition and ventilation setup.

Fig. 5.39 Façade composition, ignition, and initial ventilation setup for Case 4

(a) Ground Floor (b) First Floor (c) Second Floor

Fig. 5.40 Fire loads used in Case 4

at FF and SF. The setup of the GF compartment is shown in Fig. 5.39b and was used as the ignition source to simulate an external fire scenario for the façade. Two panes of the façade at GF were opened to provide initial ventilation, corresponding to an initial ventilation factor of $F_v = 0.01\ m^{0.5}$. No initial ventilation was provided at FF and SF. As the fire progressed and the façade panels failed, ventilation factor increased.

Fire scenarios of this experiment were developed using realistic furnishings, as shown in Fig. 5.40 with a quantitative summary of the fire loads provided in Table 5.9 (Table 5.5).

5.6.1 Behavior of Fire and Façade with Timeline of Events

5.6.1.1 Event 1: Pre-checks (−5 min)

Pre-check routines were carried out on all the systems.

Table 5.5 Quantitative summary of fire loads of Case 4

S. No.	Item	Material	Quantity	Fire load (MJ)
Ground floor				
1	Curtains/sheets	Cotton	–	51.9
2	Crib	Pine wood	2	2305.4
	Total			2357.3
First floor				
1	Curtains/sheets	Cotton	–	51.9
2	Crib	Pine wood	2	2305.4
	Total			2357.3
Second floor				
1	Curtains/sheets	Cotton	–	51.9
2	Crib	Pine wood	1	1720.9
	Total			1772.8

5.6.1.2 Event 2: Ignition (0 min)

A tray filled with gasoline was placed beneath the wood crib placed at GF, which was used to provide the initial ignition.

5.6.1.3 Event 3 (1 min)

Flames from the wood crib reached the ceiling level at the GF, and high amount of smoke was observed to come out of the compartment. No smoke obscuration was observed at the FF and SF. Maximum temperature at GF reached 135 °C, while that at FF and SF was 25 °C. Views from outside the building and inside the GF are shown in Fig. 5.41. GF internal camera failed due to heat shortly after this.

5.6.1.4 Event 4 (4 min)

GF temperatures increased rapidly to 625 °C due to the large initial ventilation that was provided. FF and SF were completely obscured with smoke due to the leakages through the unrated firestop. The glass panels showed a different failure mechanism in this experiment due to a more restrained condition provided by the masonry spandrel, as opposed to a more compliant condition which prevailed in a large aluminum frame in the other experiments. Glass panels showed typical thermal bowing as they bent toward inside the compartment where the temperatures were higher and subsequently shattered after significant bending took place. The cladding frame also showed a similar bending phenomena due to the end restraints from the vertical supports of the building.

(a) View from outside. (b) View inside GF.

Fig. 5.41 Views at event 3

5.6.1.5 Event 5 (5.75 min)

Constant impingement of the flames at the firestop region from the GF caused failure of the unrated firestop which was provided and fire reached the FF through the firestop gap. FF internal camera failed and SF was completely obscured with smoke.

5.6.1.6 Event 6 (6.5 min)

Shattering of the glass panels enhanced ventilation at the GF and the temperature rose to 760 °C. The wood crib placed at the FF caught fire due to the internal passage of hot gases and flames from the GF. Due to the onset of combustion of the wood crib, the firestop between the FF and SF also failed and hot gases began to reach the SF. The wood crib placed at SF did not yet catch fire. However, internal camera of SF failed shortly after this. The views around this event from inside SF are shown in Fig. 5.42. Movement of the SF curtain due to ingress of of hot gases from FF can be observed before the curtain catches fire.

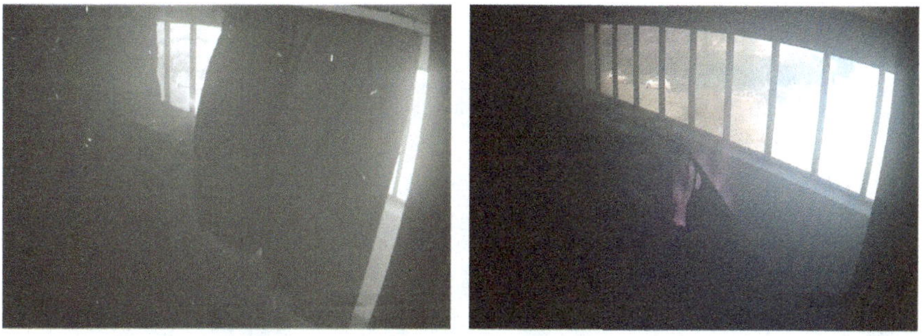

(a) View inside SF. (b) View inside SF.

Fig. 5.42 Views around event 6

5.6.1.7 Event 7 (10.93 min)

Flashover occurred at the GF with maximum temperature reaching 850 °C, while the temperature at FF was about 350 °C. Glass panels at the FF began to exhibit a similar failure mechanism as that of GF (i.e., inward bending and in-place shattering). This intensified the fire at FF and the temperature reached 660 °C. Subsequently, the fuel at the GF was consumed thereby decreasing the GF temperatures.

5.6.1.8 Event 8 (13.08 min)

Flashover occurred at the FF with maximum temperature reaching 845 °C, whereas at the GF, temperature reduced to about 380 °C. Post-flashover, the fuel at FF began to diminish. The wood crib placed at the SF had caught fire. Cement boards at all floors were intact thereby protecting beams and slab to direct exposure of heat (Fig. 5.43).

5.6.1.9 Event 9 (24 min)

Temperature reached 750 °C at the SF and the firefighting operations were started.

5.6.2 Time-Temperature Data

The maximum temperatures recorded at all floor levels during the experiment are shown in Fig. 5.44. It can be observed that GF, FF, and SF follow an almost identical temperature-time behavior except for the initial delay which corresponds to the time it take for fire to breach the unrated firestop. This highlights the importance of proper treatment of the safing area (Fig. 5.45).

(a) Event 5. (b) Event 6. (c) Event 8.

Fig. 5.43 Views from the outside at events 5, 6, 8

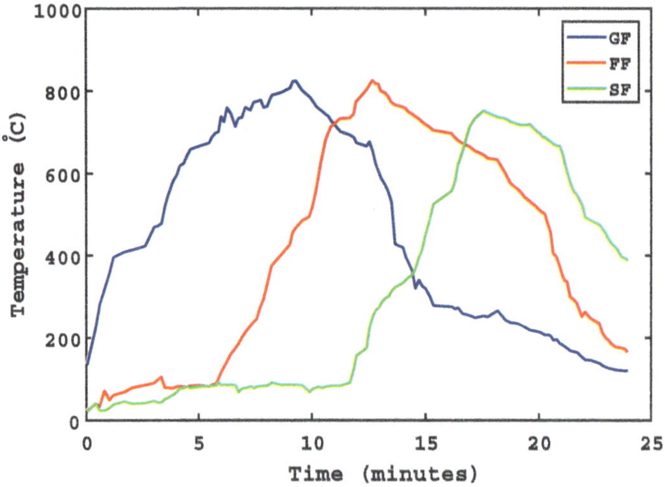

Fig. 5.44 Maximum temperatures observed at different floor levels in Case 4. Fire at all floor levels behaved identically with a time lag—the time that it took for the firestop to fail

5.7 Case 5: Glass System with Rated Firestop

This experiment utilized the same configurations as in Case 4 and was performed on 'Cabin 3' side of the fire building. The only difference was the kind of firestop. In this case, a rated firestop system comprising of compressed mineral wool and a smoke sealing joint spray was utilized in the safing area. It was expected that such an arrangement will hinder passage of hot gases and smoke to the upper stories.

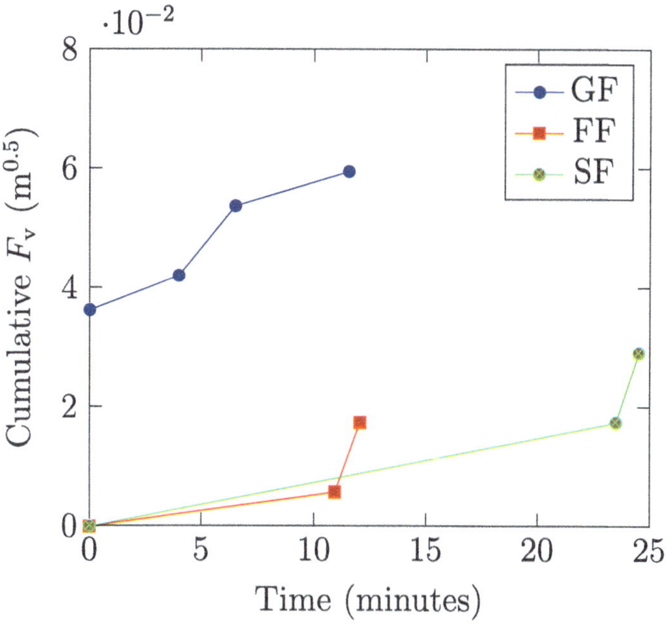

Fig. 5.45 Opening factor versus time for Case 4

5.7.1 Behavior of Fire and Façade with Timeline of Events

5.7.1.1 Event 1: Pre-checks (−5 min)

Pre-check routines were carried out on all the systems.

5.7.1.2 Event 2: Ignition (0 min)

A tray filled with gasoline was placed beneath the wood crib placed at GF, which was used to provide the initial ignition.

5.7.1.3 Event 3 (6 min)

Due to large ventilation at the GF, the GF fire grew rapidly with temperature reaching 720 °C within 6 min. A large quantity of smoke was observed to be coming out of GF, whereas almost negligible smoke obscuration and temperature rise was observed at the FF and SF, due to the good performance of the rated firestop installed between GF and FF.

5.7.1.4 Event 4 (13 min)

Flashover occurred at GF with temperatures reaching 870 °C, while the temperature at FF reached 70 °C and that at SF remained at ambient levels. The mild increase in temperature at the FF was possibly due to passage of hot gases from GF to FF through conduits used for wiring of the sensors.

5.7.1.5 Event 5 (23 min)

Fire did not propagate to FF or SF; temperature at SF remained at ambient levels while that at FF reached nearly 100 °C. None of the fire loads kept a FF caught fire. Temperature at the GF began to drop due to consumption of the fire loads. Firefighting operations were started.

5.7.2 Time-Temperature Data

Variation of maximum compartment temperatures with time is shown in Fig. 5.46. The effectiveness of a proper firestop system can be clearly observed where GF experienced a fully developed fire while temperature at FF, the floor immediately above the fire compartment, remained below 200 °C unlike Case 4, where the upper floor temperatures followed a similar trend as the fire compartment with a time delay. Temperature at SF remained at ambient levels throughout.

5.8 Case 6: Glass-ACP System with PIR Insulation

This experiment utilized a combination of an exterior insulation system (PIR boards) and ACP cladding with an air cavity in-between, as shown in Fig. 5.47. Additional masonry walls of height 1 m were constructed on the façade side at all floor levels. The remaining openings were closed with glass windows. 150 mm thick PIR boards were installed using mechanical fasteners on the masonry walls throughout the face of building. An air cavity of 250 mm was provided between the PIR boards and external ACP cladding. ACP and SGU were mounted on the external cladding frame subsequently using earlier mentioned methods.

This experiment was performed toward the Cabin 2 side of the facility. In addition to the façade opening, the insulation and cladding was also installed on the adjoining external wall. Fire scenarios of this experiment was developed using realistic furnishings, as shown in Fig. 5.48 with a quantitative summary of the fire loads provided in Table 5.6.

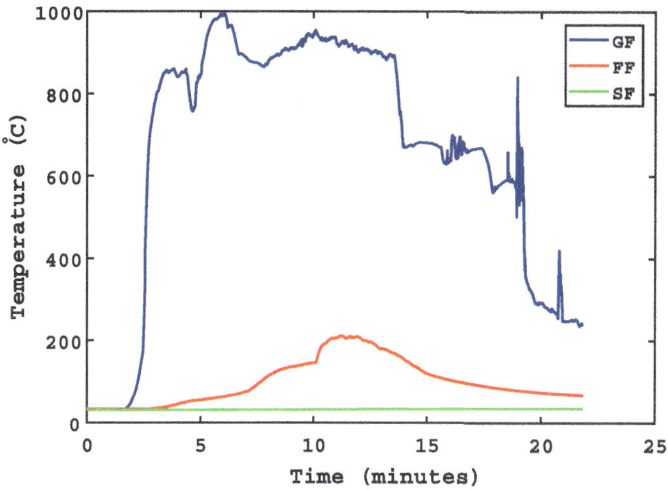

Fig. 5.46 Maximum temperatures observed at different floor levels in Case 5. Significantly lower temperatures at FF level when the fire at GF was fully developed indicated the effectiveness of appropriately designed and installed firestop system

(a) View of installation. (b) Plan of installation.

Fig. 5.47 Installation of PIR for experiment of Case 6. Yellow panels are the PIR insulation boards. Cladding frame was installed on top of the insulation boards with an air gap. Façade panels were installed on the cladding frame

(a) Ground Floor (b) First Floor (c) Second Floor

Fig. 5.48 Fire loads used in Case 6

Table 5.6 Quantitative summary of fire loads of Case 6

S. No.	Item	Material	Quantity	Fire load (MJ)
Ground floor				
1	Television	Polypropylene	1	580.8
2	Table	Wood	2	844.8
3	TV cabinet	Wood	1	668.8
4	Curtains/sheets	Cotton	–	126.4
5	Crib	Pine wood	2	5168.1
6	Sofa	Wood, polyester	1	3401.2
7	Carpet	Cotton	1	474.0
8	Newspaper, magazines	Cellulose	–	66.0
	Total			11330.1
First floor				
1	Chest	Wood	1	563.2
2	Table	Wood	1	404.8
3	Sofa	Wood, upholstery	3	3361.6
4	Curtains/sheets	Cotton	1	126.4
5	Crib	Pine wood	1	5082.1
6	Carpet	Cotton	1	474.0
7	Newspaper, magazines	Cellulose	–	49.5
	Total			10061.6
Second floor				
1	Crib	Pine wood	1	5082.1
2	Table	Wood	1	686.4
3	Chair	Wood	4	1073.6
4	Cabinet	Wood	1	422.4
5	Plates, cups	Polypropylene	–	58.1
6	Wall fixture	Wood	1	352.0
7	Curtains, sheets	Cotton	–	126.4
8	Carpet	Cotton	–	474.0
	Total			8274.7

5.8.1 Behavior of Fire and Façade with Timeline of Events

5.8.1.1 Event 1: Pre-checks (−5 min)

Pre-check routines were carried out on all the systems.

5.8.1.2 Event 2: Ignition (0 min)

The gasoline-filled pan placed beneath the wood cribs was ignited manually. In this experiment, two simultaneous ignitions were carried out—one near the small crib, one near the large crib.

5.8.1.3 Event 3 (4.17 min)

After the initial smoldering of the fire within the GF, the size of the fire began to grow as other combustible items of GF caught fire with temperature reaching 400 °C. At this stage, one of glass windows broke and the fire entered the air gap (cavity) between the thermal insulation and the façade.

5.8.1.4 Event 4 (5.05 min)

Temperature within the cavity reached around 1022 °C indicating participation of PIR in the fire and activation of the chimney effect within the cavity due to the lack of firestops within the gap. It is to be noted that due to thermal comfort considerations, discussed earlier, typically such cavities are kept free by design so that hot air can rise up throughout the building and cold air can enter from the bottom of the façade. Due to the chimney effect, the fire ignited the FF façade with temperature near the façade reaching 550 °C.

5.8.1.5 Event 5 (6.33 min)

Glass panels of the façade began to fail in intact form at all floor levels due to high temperatures which caused excessive deformations of the cladding frame as well as partial melting. The ACP sheets also dislodged from the cladding due to accelerated growth of fire because of PIR boards. ACP sheets initiated a significant secondary fire at the GF outside the building. Although most of the façade panels failed by this time, the ingress of fire did not happen inside the FF or the SF compartments. This was due to two factors: First was a general unfavorable wind velocity (i.e., the façade face was the leeward side during the experiment), and the other was a

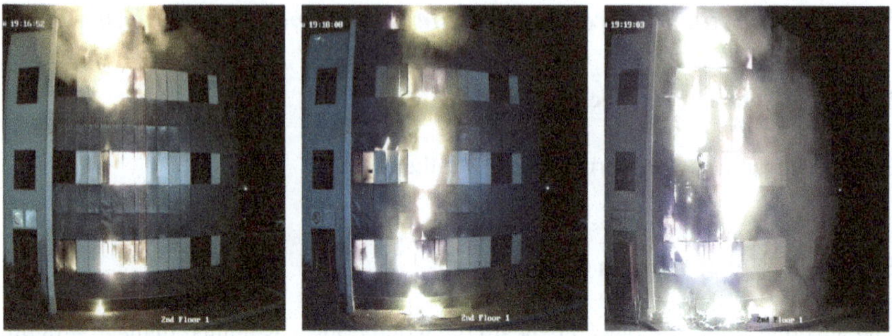

(a) Event 4. (b) Event 5. (c) Event 6.

Fig. 5.49 Views from outside at events 4, 5, 6

significant upward air current due to the chimney effect, which prevented hot gases from ingress inside the compartments in a significant way and carried them primarily upward (Fig. 5.49).

5.8.1.6 Event 6 (7.25 min)

Flashover occurred at the GF. The nearby PIR boards were consumed and mostly charred, which reduced the temperature within the cavity. Certain items kept very close to the façade face at FF caught fire while nothing caught fire at the SF.

5.8.1.7 Event 7 (9.81 min)

Due to the strong chimney effect, interaction of façade fire with compartments was not significant. At this stage, constituents of the compartments (FF and SF) were predominantly unburned as the façade fire began to diminish due to unavailability of the façade or insulation material near the compartment. A significant secondary fire was initiated outside the building due to failure of combustible façade components. This phenomenon is illustrated in Fig. 5.50 where the strong upward air draft prevents hot gases from entering the compartment. One more reason for less interaction was the wind direction on the day of experiment, which was from far end of the compartment toward the façade that further inhibited the flow of hot gases from façade to the compartment.

5.8.1.8 Event 8 (10.5 min)

Fire intensity at the GF reduced due to consumption of the fuel. Fire did not spread within the FF and remained localized to the items kept near the façade, which subse-

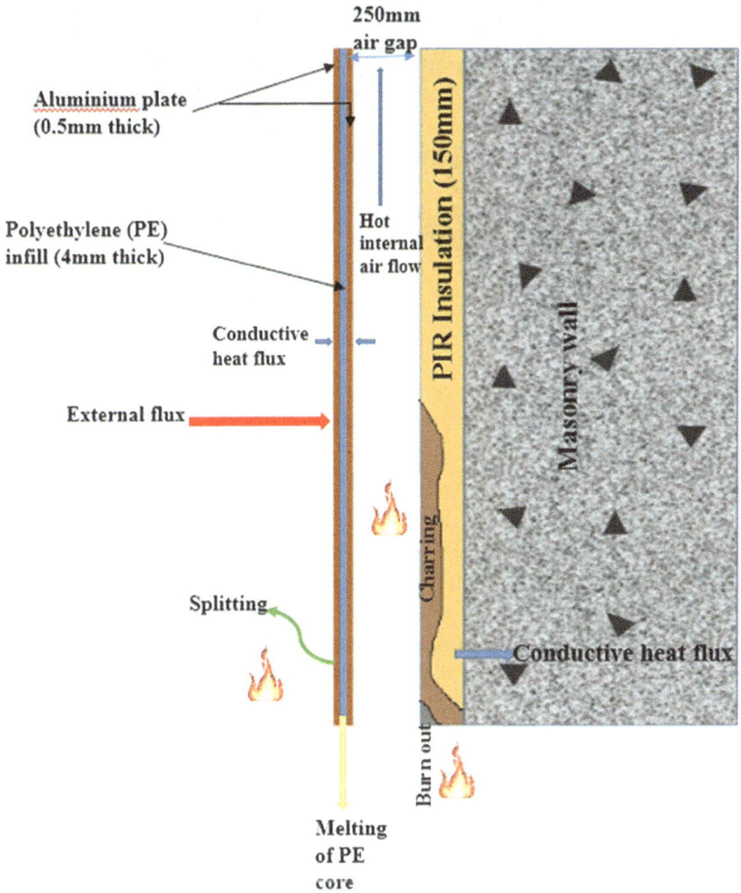

Fig. 5.50 Schematic of PIR insulation and ACP cladding used in Case 6. Once fire reaches inside the cavity, façade panels are exposed to fire from both inside and outside. Further, the chimney effect significantly intensifies the speed of fire spread

quently died on its own. PIR participated in fire at all levels and eventually became charred. The fire did not penetrate inside the SF compartment. Transmission of fire internally through the air cavity and PIR boards to the top-right corner of the building can be observed. This portion of the building did not participate in the fire until now due to an opposing wind direction. However, once the favorably positioned PIR boards were exhausted and ACP sheets dislodged, fire propagated toward the right side of the building (Fig. 5.51).

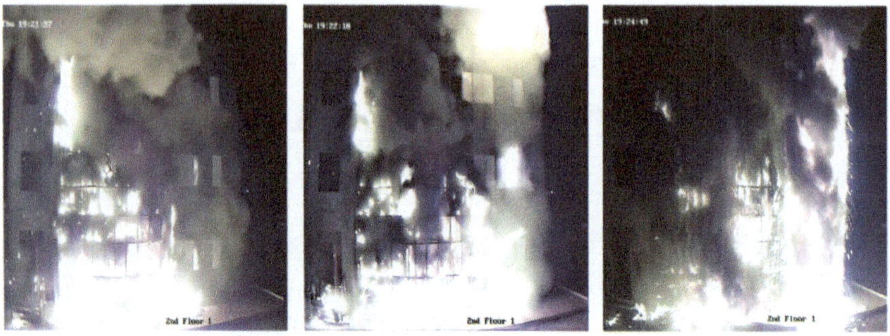

(a) Event 7. (b) Event 8. (c) Event 9.

Fig. 5.51 Views from outside at events 7, 8, 9

5.8.1.9 Event 9 (13 min)

The compartment fires of GF and FF as well as the façade fires were diminishing due to unavailability of additional fuel. Right side of the building showed significant fire contribution at this stage due to the horizontal travel of fire through the air cavity and PIR boards. A sizable secondary fire was also developed in front of the right portion of the building. Firefighting operations were initiated subsequently.

5.8.2 Time-Temperature Data

Maximum temperatures observed within the compartments are shown in Fig. 5.52. These temperatures were extracted from the thermal images as the DAQ data was lost toward the end of the experiment and could not be retrieved.

5.9 Case 7: Glass-ACP System with PIR Insulation and Façade Sprinkler

This experiment utilized the same design of exterior insulation (PIR) and ACP cladding as was used in Case 6 (see Sect. 5.8). There were two main differences between Case 6 and Case 7. First, an outside ignition source (similar to Case 1) was used. Second, a façade sprinkler system was installed in this experiment to assess the efficacy of such a system. The sprinkler system was not kept in auto mode as it was desired for the fire to grow substantially before triggering the sprinkler system manually. In real fires where automatic sprinkler systems are employed, an early triggering of the sprinkler system can be expected. This experiment was performed on Cabin 1 side of the fire building.

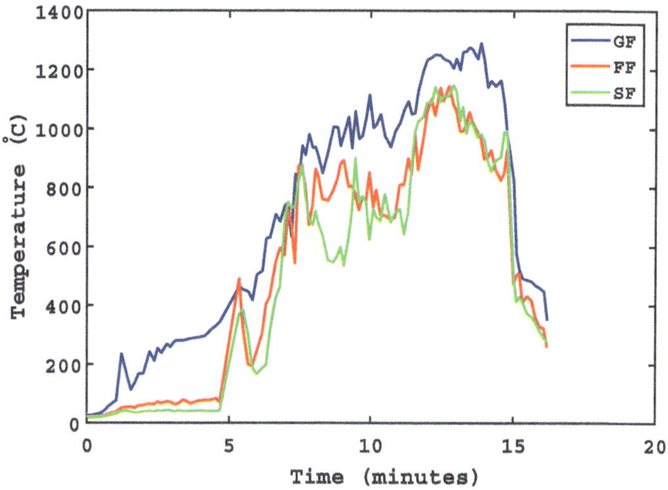

Fig. 5.52 Maximum temperatures observed at different floor levels near the façade in Case 6. Once the fire reached inside the cavity, façade of all floor levels behaved almost identically in terms of temperature and fire response

(a) View 1. (b) View 2

Fig. 5.53 Views of installation of inside sprinkler system. CPVC pipes supported on MS box pipes were used

Views of the inner sprinkler system are shown in Fig. 5.53. Sprinklers were installed on a CPVC pipe system supported on MS vertical posts. Figure 5.54 shows installation of the outside sprinkler system. Outer sprinklers on the façade surface were installed on metal pipes. The connection entailed two outlets, one for the inside sprinklers and one for the outside ones, branching out of a single input of pressurized water line.

Fire scenarios of this experiment were developed using realistic furnishings, as shown in Fig. 5.55 with a quantitative summary of the fire loads provided in Table 5.7.

Fire scenarios of this experiment were developed using realistic furnishings, as shown in Fig. 5.55 with a quantitative summary of the fire loads provided in Table 5.7.

(a) Sprinkler connection. (b) Outside view.

Fig. 5.54 Views of installation sprinkler system and water connection. Red-colored pipe was installed at the outer surface of façade while orange-colored pipe was installed from inside the compartments with sprinklers focused toward the façade

(a) Ground Floor (b) First Floor (c) Second Floor

Fig. 5.55 Fire loads used in case 7

5.9.1 Behavior of Fire and Façade with Timeline of Events

5.9.1.1 Event 1: Pre-checks (−5 min)

Pre-check routines were carried out on all the systems.

5.9.1.2 Event 2: Ignition (0 min)

The outer wood crib was ignited using a gasoline filled pan.

Table 5.7 Quantitative summary of fire loads of Case 7

S. No.	Item	Material	Quantity	Fire load (MJ)
Ground floor				
1	Printer	Polymer	1	120.3
2	Table	Wood	1	440.0
3	Small table	Wood	1	135.0
4	Curtains/sheets	Cotton	–	126.4
5	Crib	Pine wood	2	1722.2
6	Sofa	Wood, Polyester	3	13530
7	Carpet	Cotton	1	474.0
8	Newspaper, magazines	Cellulose	–	50.4
9	Chairs	Wood, upholstery	2	126.0
10	Cushion	Polyester	5	715.0
11	Small chest	Wood	1	563.2
	Total			18002.5
First floor				
1	Desk	Wood	1	1589.0
2	Single bed	Wood	1	525.0
3	Cupboard	1	874.0	
4	Curtains/sheets	Cotton	1	126.4
5	Crib	Pine wood	2	3444.4
6	Carpet	Cotton	1	474.0
	Total			7032.8
Second floor				
1	Crib	Pine wood	2	3444.4
2	Chair	Wood	2	536.8
3	Curtains, sheets	Cotton	–	126.4
4	Carpet	Cotton	–	474.0
	Total			4581.2

5.9.1.3 Event 3 (3.33 min)

Fire of the external wood crib was fully developed and temperatures of thermocouples near the façade began to rise.

5.9.1.4 Event 4 (3.60 min)

One glass panel at the FF failed. Temperature within the air cavity was greater than 150 °C; cavity temperatures were greater at SF compared to that of GF due to the fact that fire reaching the cavity was from the wood crib and not due to the ignition of insulation material.

(a) Event 3. (b) Event 5. (c) Event 6.

Fig. 5.56 Views from outside at events 4, 5, 6

5.9.1.5 Event 5 (3.65 min)

Insulation material at SF level caught fire and an independent flame within the air cavity was observed. A sharp rate of temperature rise within the cavity was observed (Fig. 5.56).

5.9.1.6 Event 6 (3.98 min)

Fire spread laterally across the insulation material at SF within the air cavity. ACP and PIR at GF began to catch fire due to prolonged radiant exposure to the external wood crib fire. Temperatures within the air cavity were greater than 500 °C. Temperatures within the compartments were at ambient levels. ACP core material began to melt and drip around this time due to heat exposure and was observed in the form of burning embers falling from the façade.

5.9.1.7 Event 7 (4.72 min)

Substantial portion of the façade was under fire and many panels failed by this time. ACP panels that fell on the ground initiated a secondary fire. Temperature within the cavity at SF reached beyond 950 °C while temperature inside the SF compartment was in excess of 600 °C. Maximum temperature at GF and FF compartment was 40 and 350 °C, respectively.

5.9.1.8 Event 8 (5.03 min)

Many portions of the façade cladding frame failed by this time (large fragments fell down due to failure at joints or melting of small portions). Secondary fire grew in

(a) Event 7. (b) Event 8.

Fig. 5.57 Views from outside at event 7, 8

size considerably. Temperature within the cavity reached 1000 °C. Maximum compartment temperatures were 45, 540, and 630 °C at GF, FF and SF, respectively (Fig. 5.57).

5.9.1.9 Event 9 (5.10 min)

Sprinkler system was switched on, and a declining trend was initiated in the measured temperatures.

5.9.1.10 Event 10 (5.93 min)

Fire pertaining to façade was extinguished by this time and temperatures within the air cavity dropped below 200 °C. Since the last event, the building was engulfed in a significant amount of vapor due to the water sprayed by sprinklers. Since GF compartment temperatures never reached above 50 °C, no compartment fire was initiated; a gradual drop in compartment temperatures at FF and SF level was observed (Fig. 5.58).

5.9.1.11 Event 11 (9.00 min)

Most thermocouples were showing ambient temperatures as façade fire was extinguished by the sprinkler system. Fire of the external wood crib and contents of the SF was not extinguished by the façade sprinkler system. External firefighting operations were initiated at this time to end the experiment.

(a) Event 10. (b) Event 11.

Fig. 5.58 Views from outside at event 10, 11

5.9.2 Time-Temperature Data

The maximum temperatures recorded within compartments and the insulation-cladding air cavity are shown in Figs. 5.59 and 5.60, respectively. It can be observed that cavity temperatures increased sharply once fire reached the PIR boards; compartment temperatures at FF and SF followed suite but were lower in magnitude. Temperatures at GF (both for cavity and compartment) remained relatively lower. Façade temperatures registered a steep decline after the initiation of the sprinkler system. The sprinkler system was not effective in controlling compartment fire; SF temperatures increased again after sprinkler stopped due to a sustained fire within the compartment.

5.10 Discussions

This section presents discussions on various aspects of the case studies, including fire behavior and failure mechanisms of different components.

5.10.1 Fire Behavior

Fire behavior can be considered under two regimes. First, if it were a regular room (i.e., without façade), the expected time-temperature behavior can be predicted by the Eurocode parametric fire curves. In this scenario, two conditions can be considered to give a lower and an upper bound on the fire severity, as shown in Fig. 5.61. If the

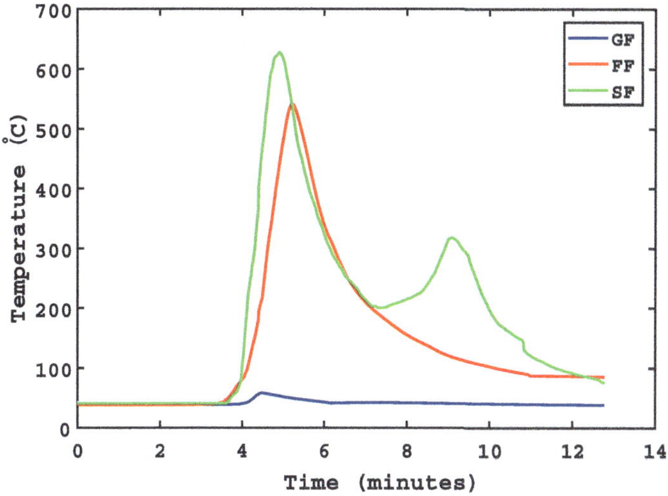

Fig. 5.59 Maximum temperatures observed at different floor levels in Case 7. Gradual drop in temperature was observed; at SF the fire was not completely extinguished, and the intensity increased after sprinkler system was switched off. This indicated the ineffectiveness of a façade sprinkler in controlling a room content fire

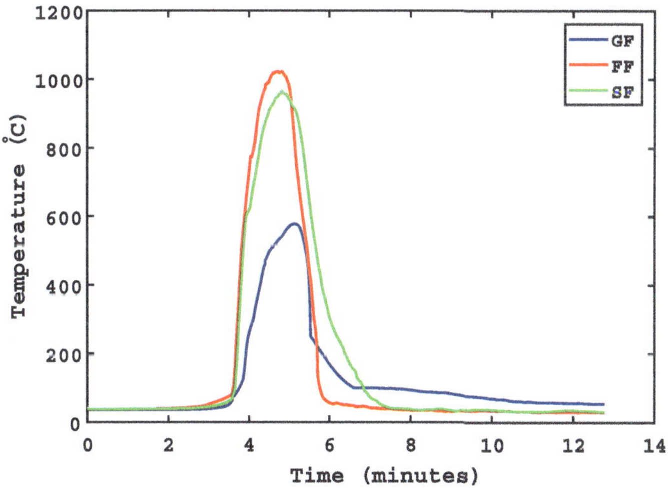

Fig. 5.60 Maximum temperatures observed at insulation-façade gap in Case 7. Sudden drop in temperature was observed after the sprinkler system was triggered. This indicated effectiveness of façade sprinklers in controlling façade fires

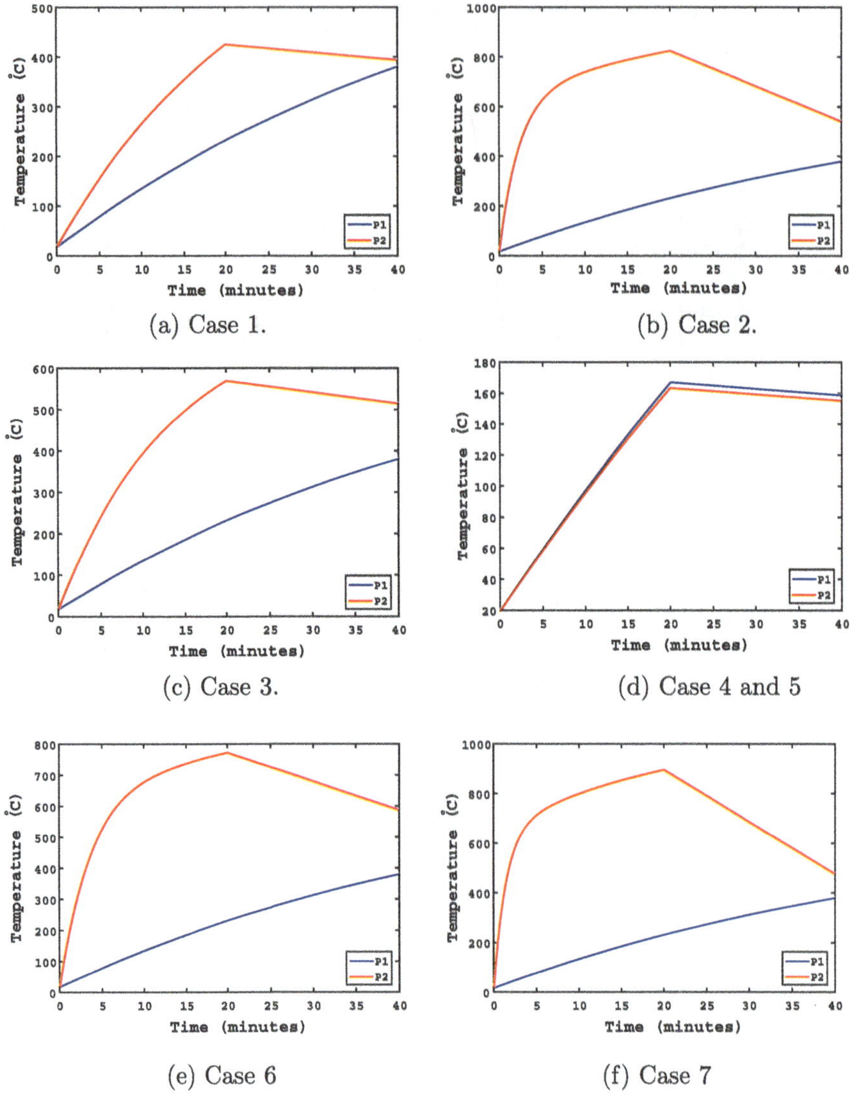

(a) Case 1. (b) Case 2.

(c) Case 3. (d) Case 4 and 5

(e) Case 6 (f) Case 7

Fig. 5.61 Parametric fire curves for different cases under initial and extreme conditions. 'P1' denotes the parametric fire curve if the initial opening condition prevailed throughout; 'P2' denotes the parametric fire curve if the entire façade failed

initial ventilation condition prevailed (i.e., the façade did not fail), it will correspond to a lower bound on the fire severity (shown by P1), whereas if the entire façade failed in the beginning, it will correspond to an upper bound on the fire severity (shown by P2). It can be noted that the Eurocode parametric fire curves consider the effects of the total fire load, ventilation conditions, and the thermal inertia of the wall, ceiling, and floor. Further, the graphs shown here are for GF only. If one compares these parametric fire curves to the observed maximum time-temperature behavior of the fires in the experiments, it becomes clear that the usual calculation of the parametric fire will is not effective here, primarily due to the contribution of the combustible façade components and the interaction between the compartment fire and façade fire (recall that independent fires in the compartment and the façade were observed in several cases).

This leads to consideration of a second regime of fire behavior—to incorporate the interaction between the compartment and the façade fires. As pointed out in Chapter 4, this is an active area of work and involves use of high-fidelity computer models. A large number of studies are required before simplified procedures (akin to the Eurocode parametric fires) can be developed.

Further, one can consider Q_{st} for each of these cases from (4.10), which will be an empirical upper bound of the heat release rate, as discussed earlier. Interestingly, it depends only on the window dimensions and not on the fire load of a compartment. Similar to the parametric curves, Q_{st} can be calculated for two regimes: with initial ventilation conditions and with full façade opening. Using Q_{st} and the fire load of a compartment, it is possible to obtain an estimate of the total time of burn, t_{burn} (assuming Q_{st} to be heat release rate in an average sense), as shown in Table 5.8. It can be expected that the observed time of burn will be within the bounds given by Table 5.8. While it is true for certain cases, there are anomalies too. The excess heat flux to which the façade panels were subjected to was found to be in as high as 140kW/m^2 (Srivastava 2020). Regulatory test procedures usually consider up to 100kW/m^2 levels of exposure conditions, as discussed in Chapter 4.

5.10.2 Façade Failure Mechanisms

This section discusses the key features of failure of different components of the façade system.

5.10.2.1 Global Failure Features

Some global failure patterns and features are shown in Figs. 5.62, 5.63, 5.64, and 5.65. External ignition was found to cause more severe fire compared to internal ignition with other parameters being the same (Case 1 vs. Case 2). This is due to a greater exposure of the façade system to flames in case of external ignition (see Fig. 4.2 for a conceptual mechanism of fire spread). Common features observed across all the

Table 5.8 Heat release rate and time of burn calculations for GF. $A_T = 90\,m^2$ for all cases

Parameter	Case 1	Case 2	Case 3	Case 4	Case 5	Case 6	Case 7
Fire load [GJ]	4.63	14.05	6.16	2.34	2.34	11.36	18.00
For initial ventilation condition							
W_V [m]	2.4	2.4	2.4	1.0	1.0	2.4	2.4
H_V [m]	1.2	1.2	1.2	2.2	2.2	1.2	1.2
A_V [m^2]	0.96	0.96	0.96	2.20	2.20	0.96	0.96
F_V (4.4) [$m^{0.5}$]	0.012	0.012	0.012	0.034	0.034	0.01	0.01
n (4.16)	4	4	4	0.91	0.91	4	4
Q_{st} (4.10)[MW]	1.58	1.58	1.58	4.89	4.89	1.58	1.58
t_{burn} [min]	48.93	148.48	65.13	8.03	8.03	119.71	190.20
For full façade failure condition							
W_V [m]	4.0	4.0	4.0	5.0	5.0	4.0	4.0
H_V [m]	2.4	2.4	2.4	1.4	1.4	2.4	2.4
A_V [m^2]	9.6	9.6	9.6	7.0	7.0	9.6	9.6
F_V (4.4) [$m^{0.5}$]	0.165	0.165	0.165	0.092	0.092	0.165	0.165
n (4.16)	3.33	3.33	3.33	7.14	7.14	3.33	3.33
Q_{st} (4.10)[MW]	22.31	22.31	22.31	16.27	16.27	22.31	22.31
t_{burn} [min]	3.45	10.50	4.61	2.42	2.42	8.46	13.45

experiments included— failure of cladding frame due to melting and large deformations; accumulation of debris near the GF as a result of falling façade materials which also caused secondary fires; and failure of safing area firestop due to hot movements of façade and permanent deformations of the cladding frame.

5.10.2.2 Cladding Frame Failure

Cladding frame failure initiated due to exposure to high temperatures as aluminum quickly loses half of its strength in the temperature range of 250–300 °C and melts at around 660 °C (Summers 2015). Melting of the cladding frame can be observed in the earlier presented photographs. Reduced strength of aluminum also caused dislodging of large segments of cladding frame in certain cases (specially Case 1 and Case 6). Additionally, many typical modes of failure of metal structures at high temperature were observed, as shown in Fig. 5.66.

5.10.3 Firestop Behavior

Safing firestop installations typically failed early due to hot movement of the cladding frame which allowed passage of smoke to the upper compartments. The use of a rated

Fig. 5.62 General features of façade failure for case 1. The following can be observed: large extent of façade failure due to external ignition source; interaction between façade and room contents fire (particularly at FF)

Fig. 5.63 General features of façade failure for Case 2. The following can be observed: significantly smaller extent of façade failure compared to Case 1 (see Fig. 5.62); flame spread through combustible façade panels at the central portion; different stages of ACP in its participation in such a fire

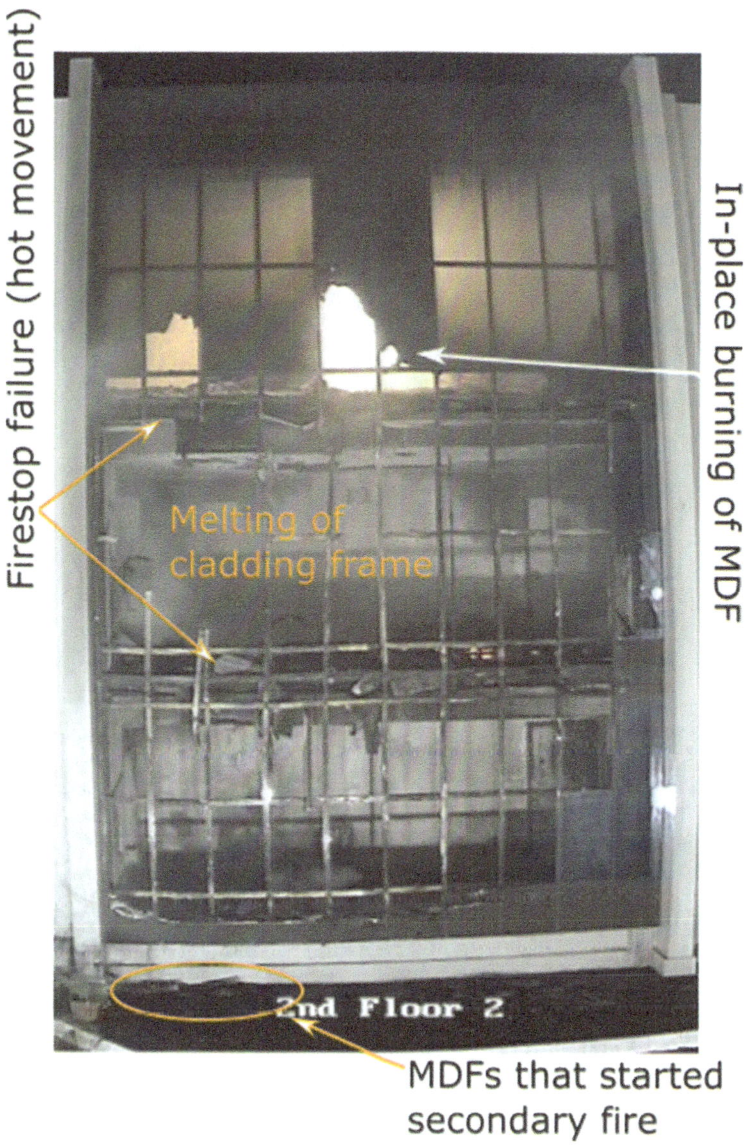

Fig. 5.64 General features of façade failure for Case 3. The following can be observed: large deformations and dislodging of cladding frame; in-place burning of MDF; debris at GF of façade panels and cladding frame; lower intensity failure of façade of upper floor due to the use of MDF

Fig. 5.65 General features of façade failure for Case 6. The following can be observed: charring of PIR boards; large permanent deformations in cladding frame; dislodging of cladding frame; failure of all façade panels (ACP) and glass; large amount of debris of fallen façade panels and cladding frame

(a) Mode 1. (b) Mode 2. (c) Mode 3.

Fig. 5.66 Failure modes of cladding frame. **a** Pulling action by the transom on the mullion due to catenary action of the transom caused due to reduced strength and thermal expansion at high temperature. **b** Dislodging of frame fragments due to melting of an entire cross section of aluminum at high temperature. **c** Local buckling of mullion due to compressive forces from the transom due to their thermal expansion; due to lower strength and modulus at high temperatures, the buckling strength of frame sections decreases

firestop system in Case 5 was the only exception to this. In all cases except 4 and 5, spandrel area was made of combustible panels (ACP and MDF). In later stages of fire, safing mineral wool of the firestop system was completely exposed due to dislodging of the cladding frame and spandrel panels. If the spandrel panels were non-combustible, or protected from fire from the outside also, the performance could have been better. Taking Case 2, for instance, it can be observed for the FF/SF firestop (Fig. 5.63) that a central portion where the spandrel area ACP burnt, the compression of mineral wool was completely lost whereas in areas toward the building edges, spandrel area ACPs experienced hot movements due to exposure to external flames of the façade fire (Fig. 5.67).

In Case 5, the safing firestop worked well and did not allow passage of hot gases due to the spandrel area being non-combustible (and no hot movements) as it was made of masonry. Figure 5.68 shows the photographs of the electrical conduit and pipe firestop systems that were installed during the experiment. Both the systems worked well in the expected manner by closing the gaps through the expansion of the firestop material during fire (before fire photographs of these installations can be seen in Fig. 3.5).

5.10.4 Panel Failure

As discussed earlier, glass panels showed two types of failure mechanisms. In compliant installations (full face cladding frames), glass panes dislodged from the cladding frame due to (i) burning of the pressure tape and silicon sealant, and (ii) large defor-

Fig. 5.67 Failure of firestop at the spandrel level. Local deformation in the cladding frame can be observed which loosen the firestop filling and cause failure of the firestop. The deformations are primarily due to hot movement of the cladding frame during fire

(a) Conduit firestop after fire.

(b) Pipe firestop after fire.

Fig. 5.68 Firestop systems after a fire experiment. **a** Firestop material expanded to fill the cavity (black colored material). **b** CPVC pipe melted during fire and firestop material expanded to seal the cavity

mations of the cladding frame. In more restraining installations (cases with masonry spandrel), glass panes bent in-place and shattered after undergoing large deformations.

ACP failed initially due to expansion of the aluminum sheets and failure of the adhesive layer that holds the polymeric core between aluminum sheets. Once the edges of ACP were exposed to temperatures greater than 200 °C, the polymer caught fire and began to burn leaving a black residue (typical of such polymers) on the aluminum sheets. As can be observed from the earlier photographs, some ACP sheets burnt in-place, while others dislodged from the cladding frame and started secondary fires at the GF outside the building. Well before ACP sheets showed significant failure, the polymeric core began to drip and aided the local combustion by infusing the vicinity with combustible volatile matter. Such failures of ACP are well documented in the past, and there are a newer variety of fire-resistant ACPs available in the market which claim better fire performance and tend to alleviate some of the aforementioned issues.

MDF, being a wood-based panel, is inherently combustible but showed much better behavior as compared to ACP. MDF panels charred for a significant duration and held their strength. Charring is known to moderate the surface combustion rate in wood. This effect was so prevalent that in Case 3, fire could not propagate from one story to the other through façade, whereas in all the ACP façades, there was a significant role of façade in the fire growth.

PIR boards, being petroleum based, are highly combustible and are known to produce toxic gases during combustion. They significantly enhanced the fire spread in Case 6 and demonstrated a charred burning behavior.

5.10.5 Sprinkler Performance

Sprinkler system, used in Case 7, proved to be very effective in reducing the fire size. Façade and compartment temperatures reduced significantly within one minute of sprinkler activation. Automatic activation was deliberately not utilized to assess the efficacy of the system. It can be expected that if the sprinkler system is set to activate around 60 °C, its usual activation temperature, the fire may not travel through the façade at all. There is one caveat though, specially if the ignition sources is outside the building, as it was in Case 7. External sprinklers, being focused toward the façade surface, failed to extinguish the external wood crib fire. Thus, in case of an external ignition source, there exists a possibility that water of sprinkler tank may finish without extinguishing the external fire and façade materials may ignite subsequently. However, wetting of the façade materials is likely to significantly reduce the risk of fire spread through façade in any case. Another point to note is that façade sprinkler system was also ineffective in extinguishing a well-developed compartment fire (in Case 7); thus, a façade sprinkler system, if used in addition to a usual sprinkler system, may prove to be the most effective. Lastly, there was practically no difference in the use of metal versus CPVC pipes for installing the sprinkler. Since they are water filled, CPVC pipes are able to sustain the heat from fire and fulfill the intended purpose. Due to the ease of workmanship and negligible corrosion potential, CPVC pipes may be preferred in such installations.

5.10.6 Passive Fire Protection

Two types of passive fire protection were used—masonry and board. Both worked with varying degrees of success under different scenarios with masonry protection being more robust.

5.10.6.1 Masonry Walls

Masonry walls performed well during the fire experiments in providing the necessary fire protection (to structural steel columns) and as partition walls between different compartments. The only issue was spalling of the plaster which necessitated repairs before any new experiment was to be performed. Varying degrees of spalling of masonry walls and fire protection of structural columns are shown in Fig. 5.69.

5.10.6.2 Board Protection

Board failure is shown in Fig. 5.70. While the boards themselves were able to withstand at least 30min of fire exposure (due to the thickness; also tested separately at component level using a gas burner), their failure was heavily depending on the fixing detail. Different levels of failure were observed depending on the localized fire intensity and variation in fixing details (a workmanship issue). In certain cases, there was no failure of the boards at all. Other degrees of failure are shown in Fig. 5.70.

(a) Partial spalling of plaster. (b) Complete spalling of plaster.

Fig. 5.69 Varying degrees of spalling in the plaster of masonry walls and enclosures. **a** Partial spot-spalling where bulk of the plaster remained intact and localized spalling was observed; this is indicative of local pockets of high moisture content within the plaster. **b** Spalling of big portions of plaster exposing the brick masonry

(a) Partial failure. (b) Moderate failure. (c) Complete failure.

Fig. 5.70 Different levels of failure of the two-layer board protection provided for beams and deck slab. **a** Partial failure that involved failure of only the cement board of false ceiling while POP boards remained intact. **b** Moderate level of failure involved failure of cement as well as POP boards. **c** Complete failure of both the boards that exposed deck slab to high temperatures for a longer period of time which caused permanent thermal deformations in the deck slab

5.11 Challenges in the Experimental Program

This section summarizes and highlights the main challenges encountered during the series of experiments presented here.

5.11.1 Fire Protection

There are usually three choices when it comes to providing passive fire protection to structural systems: encasing the members (in concrete or masonry), use of board protection, and use of intumescent paints or foams. Appropriate methods are chosen depending on the design demands. Table 5.9 summarizes the key features of the three fire protection methods. Naturally, where a large number of experiments are to be performed within the same setting, masonry appears to be the best option considering its high effectiveness and low cost. However, due to the high weight, it is not suitable everywhere. While board protection was used with reasonable success during the experiments, their robustness remained a concern as it depended on the method of installation to a great extent. Unlike this, the method of construction of masonry has little effect on its effectiveness as a fire protection. Moreover, in the local context, it was much easier to find workers skilled in doing masonry work as compared to

Table 5.9 Comparison of different fire protection methods

Feature	Masonry	Board	Intumescent
Level of protection	High	Moderate	High
Robustness	High	Low	Moderate
Reusability	High	Low	Low
Weight	High	Moderate	Low
Cost	Low	Moderate	High

those adept in installing board protection. The need to remove and reinstall board protection before every new experiment adds to the overall costs of the experiment but is an unavoidable requirement.

Intumescent paint or foam is widely used for fire protection of steel buildings due to their low weight and good robustness. However, it is important to ensure good surface finish of steel before such paints can be applied (presence of rust can make them grossly ineffective as they depend on adhesion to the steel surface). In the context of these experiments, the need of reusability of the same structure and high cost of intumescent coatings (material as well as application) prevented their use to a large extent. Intumescent paint was used sparingly to protect wires going through thermocouple trees. Fire protection of beams and deck slab remains a challenge and in the current context, and the use of board protection was found to be a reasonable compromise between reusability, robustness, weight, and cost.

5.11.2 Instrumentation

Instrumentation posed varying degrees of challenges. Wiring of sensors to be placed inside fire compartments was challenging as it was to be protected from fire damage, and wire length was limited by sensor attenuation limits. Given the full scale of the test facility, it was not possible to take the DAQ ends of all the sensor wires outside the building. Thus, DAQ had to be kept inside the building. This was a major challenge and required a robust fire compartmentation to protect the DAQ during experiments. While masonry partitions performed well, minor openings and construction tolerances posed practical issues; these not only allowed passage of hot gases and smoke in some experiments, but also provided passage to water during the firefighting operations. A special metal enclosure was made to house the DAQ. It could not be air-tight to allow heat dissipation of the DAQ microprocessors; thus, it had to be oriented the right way to avoid direct exposure of the DAQ to smoke or water. Figure 5.71 shows a photograph of the DAQ and its enclosure.

Internal video cameras were usually installed at the ceiling level. Hence, their wires passed through the air cavity between cement board false ceiling and deck slab. Wires of the sensors installed at the ceiling level also utilized the same path. While this was fine for initial stages of fire, once the cement boards began to fail, these wires were exposed and data transmission to the DVR and DAQ was lost from those cameras/sensors. Wires of the lights installed inside the compartments also passed above the false ceilings and were consequently vulnerable to burning after failure of cement boards. An interesting practical consideration emerged from this issue. Sensor and camera wires used DC signals, and their short circuiting (due to burning) was handled at the DVR or DAQ level and did not cause any major disruptions beyond the loss of signal from that particular sensor. Internal lights utilized AC for their power and hence, short circuiting of wires of even one of the internal lights could potentially cause power interruption to the entire circuit. In fact, in some of the experiments, burning of one light caused outage of lights in all the compartments.

Fig. 5.71 Photograph of the DAQ within stainless steel enclosure with the provision of an openable door on one side and in-built vents to enable heat dissipation

Keeping this in consideration, power sources of lights, DVR, and DAQ were kept isolated. DAQ was powered through a generator set with no other peripherals attached to its line to ensure a dedicated power supply for data acquisition and to alleviate its dependence on the grid (which can be unreliable at times due to faults outside the premises). DVR and internal lights were powered through independent sources from the grid.

Since different systems were used to record different data DAQ for sensor data, DVR for video camera data, and independently operated thermal imaging camera—synchronization of events required careful planning. This was achieved through the use of multiple persons working as a team to signal different events during the experiments. Redundancy in observations was the key to help in this challenge.

5.11.3 Workmanship

In experiments of full scale, control of workmanship can be challenging, particularly due to the fact that majority of the manpower employed for construction, wiring, etc., lies in the unskilled category. Translation of the technical requirements and drawings in a form and language that can be understood by the workers is critical. In one of the experiments, lack of better communication caused improper fixing of the cement board false ceiling and they failed prematurely. Similarly, wiring of sensors can be challenging if traditional wiremen are employed without proper briefing.

Sequencing of events and parallel working of different construction teams was another issue. For instance, installation of the entire cement board false ceiling cannot be done in one step; it was required to leave spaces to allow wiring of sensors, cameras, and lights. Once these were installed, remaining cement boards were to be installed. These two activities were performed by different teams and close coordination was required during this time. The last activity on the false ceiling was to fill the gaps between cement boards with POP paste which would harden before the experiments. Close supervision was required during this stage as well to avoid covering of thermocouples with this paste. Use of long thermocouples was helpful in alleviating this challenge.

There were other similar activities that required close coordination between different teams, such as

- installation of floor-level sensors and placing of furniture;
- installation of firestop and façade spandrel panels;
- placing of furniture after installation of firestop (there was a tendency for workers to step on the firestop); and
- connection of sensor wires to DAQ data cards and connection of data cards to the DAQ (DAQ being a sophisticated and expensive computer was connect at the end, before beginning of the experiments).

Chapter 6
Summary and Future Directions

Façade systems have found new expressions with novel materials in modern buildings and are an important component of green building strategies. Their primary functions are to enhance aesthetics, improve energy efficiency, and provide weather protection to buildings. Conventional materials like masonry veneers, precast concrete panels, and stone cladding are being increasingly replaced by newer materials such as polymer boards and metal clad panels. These newer materials offer significant advantages over traditional materials in the following:

- Aesthetics: Modern materials enable non-traditional elevations and expressions.
- Energy efficiency: Modern materials usually have significantly lower R-values compared to the traditional ones, which provide better insulation of the building from its surroundings.
- Ease of construction: Modern materials can in general be easily prefabricated and molded in different shapes on- or off-site. Further, due to being lower in weight, provide more economical structural designs.
- Indoor environment: Use of glass allows visual connections of inside and outside of the building and enables better but controlled reach of daylight which can significantly enhance the indoor living environment.

However, in recent years, these newer façade materials have been found to adversely affect the fire safety of buildings. Façade systems of many high-rise buildings played catalytic roles in enhancing fire severity of documented fire accidents. This has primarily been due to the fact that materials with very low R-values (and high thermal capacity) are usually plastic foams which are highly combustible, as discussed in Chap. 2. Many such plastic materials (e.g., polyethylene) melt before burning and increase the local concentration of combustible vapors, which enhances the fire severity.

After the advent of use of such materials in façade systems in the 1970s, many regulatory codes and procedures have been developed in different countries that aim

G. Srivastava and P. D. Gandhi, *Performance of Combustible Façade Systems Used in Green Building Technologies Under Fire*, Springer Transactions in Civil and Environmental Engineering, https://doi.org/10.1007/978-981-16-3112-2_6

to quantify the combustibility and potential of a material to invigorate fire at different scales. Three scales where standardized testing procedures have been established are—component scale, intermediate scale, and assembly scale. Component-scale testing quantifies the basic combustion properties (pyrolysis behavior, heat release rate, heat of combustion, etc.) and behavior (smoke release rate, combustion by-product gases, time to ignition, critical temperature, critical heat flux, etc.) at material level and involves bench-scale equipment such as a thermogravimetry analyzer, cone calorimeter, and bomb calorimeter.

Assembly-scale testing rigs consider assemblies of façade systems, as they would be installed in practice, as per manufacturers' designs. These assemblies incorporate all the essential features of the end-product (including materials and procedures). Façade assemblies are subjected to standardized heating regimes, typically through calibrated wood cribs or gas burners, and assessed based on pre-defined performance criteria. Intermediate-scale testing (or subassembly testing) is in between the component and assembly scales, but can themselves involve a broad variation in terms of the assembly size and fire severity. Chapter 4 provided a detailed discussion of standardized tests of different scales and included a discussion on the rationale behind the choice and type of fire severity.

Despite all of the existing testing methodologies, major building façade fires are still prevalent and, in fact, on the rise, as discussed in Chap. 1. The primary reason for this is the large variability in the testing parameters and acceptance criteria adopted by codes of different countries. The differences in actual field conditions (in a real building) and test conditions (in a laboratory) compounded by a lack of correlation between the two also contribute toward potentially large fire risks in real installations even though a system might have performed well in laboratory testing. In a nutshell, one can say that where the performance is real (i.e., in actual buildings), it is not measured (e.g., temperature, heat flux, smoke obscuration) and where the performance can be measured (i.e., in a laboratory), it is not real (or less real). With an aim to bridge this gap, a full-scale real fire testing facility was developed at IIT Gandhinagar in collaboration with Underwriters Laboratories in 2016. This facility is a three-story structure with three test compartments at each floor (compartment plan dimensions 6×3 m and story height 3 m). It has been utilized to carry out a variety of façade fire experiments under different real fire scenarios. Chapter 5 presented in detail six case studies involving different façade systems under different fire scenarios. Key findings and limitations of these case studies are presented in subsequent sections.

6.1 Key Findings from the Case Studies

Key findings from the full-scale experiments are summarized as follows:

- Fire behavior in the full-scale experiments was found to be qualitatively similar to real building façade fires. These included—nature of fire spread from floor to floor and side to side, effect of combustible vs. non-combustible spandrel, behavior of firestop, negative effects of wall insulation and air cavity, and potential of onset of secondary fires outside the building due to dripping of melted façade materials. It was found that façade panels could be subjected to very high heat flux from the window flames (as high as 140 kW/m^2) at certain times during a fire. This is greater than the maximum incident fluxes considered by any of the existing standard testing methods (see Chap. 4).

- Opening factor plays an important role in deciding the height and nature of flames coming out of a window of a compartment. Existing test methods are based on intensive experimentation on compartment fires with openings and usually entail a fixed configuration of openings. During a real fire, as was observed in Chap. 5, the size of opening keeps on increasing due to successive failure of façade panels. This creates a dynamic situation where the opening factor (and hence, the nature of flames) keeps on changing during a fire incident. As such, correlations between window configurations and flame height/flux can become ineffective in dynamic scenarios of this kind. This was also shown through a comparison of the actual time–temperature observations vis-a-vis the Eurocode parametric fire curves developed for different opening factors for the case studies, wherein the observed behavior was in general found to be more severe compared to the parametric curves. Heat flux calculations and time of burning estimates showed a similar trend.
- Cladding frame showed typical structural fire failure mechanisms. These included pulling action by transoms on mullion (due to tensile force acting on transom), local buckling of mullion (due to compressive forces from transoms on both sides), global buckling of mullions (due to reduced material properties at high temperatures increasing the slenderness ratio), and melting (due to temperatures being greater than the melting point of aluminum).
- Failure of glass panels was found to depend on the mechanism of attachment and degree of restraint offered by the cladding frame. In more compliant frames, glass panels dislodged from the frame before shattering and fell in intact form. When more restraint was provided by the frame, glass panels shattered in-place due to thermal gradients.
- ACP sheets failed by combustion of the adhesive layer which debonded the inner polymer core with outer metal sheets, which exposed the polymer core to greater heat fluxes. Heat exposure also causes melting and pyrolysis of the polymer material which increased the local fuel vapor concentration further invigorating the fire; melted polymer initiated a secondary fire outside the building in all cases.
- MDF, being wood based, showed typical charring behavior exhibited by wood and held their strength for a longer duration. Charring also slowed the growth rate of fire. In fact, in MDF façade, fire did not propagate from one story to the other. The only issue was ingress of smoke and mildly hot gases to the upper stories due to partial failure of the firestop due to hot movement of façade at the spandrel level.
- Firestop behaved differently in different spandrel configurations. In non-combustible spandrel, the firestop system was found to be very effective in preventing passage of hot gases and fire to the upper floors. In combustible façades,

firestop was extremely vulnerable to hot movements and burnout/falling of spandrel panels which compromised the firestop. This emphasized the requirement of incorporating hot movement effects in standardized testing of firestops. Other firestop systems of through-penetration of plumbing pipes and electrical conduits were also tested in one of the experiments.

- A combination of external cladding and internal insulation with an in-between air cavity was found to have very poor fire performance. The air cavity caused a rapid spread of fire throughout the height and width of the façade system. Insulation boards (PIR) catalyzed the fire to a great extent.
- Interaction between compartment and façade fires was observed. At several instances, independent compartment and façade fires were observed. Façade fire initiated fire in upper floors and eventually led to flashover at upper floors. External ignition source was found to cause more severe fires as compared to internal ignition sources. This can be significant in the context of tall buildings, where a small façade fire at an upper floor can cause a secondary fire at the base of the building, which in turn can act as an external ignition source for rest of the building.
- Passive fire protection of different types was used for structural components of the fire building. Masonry enclosure with air gaps was found to perform well during all the experiments. Low to high levels of spalling of plaster were observed due to high temperatures, but the overall functionality of masonry protection was maintained throughout. A two-layer board protection system was found to be sensitive to installation practices and experienced moderate to large failures during experiments. In certain cases, the failure was so extreme that it exposed the structural beams and deck slab to fire; redundancy in the structural system became useful in such situations.

6.2 Advantages of Current Methods

- Generation of data in real fire scenarios which can be used for validation of computer models, development of simplified models for design, and improvement of standardized testing methods.
- Assessment of entire systems in real scenarios—although assembly-level tests consider façade assemblies, they usually do not have fire compartments beyond ground level. Thus, the effects of interaction between compartment and façade fire cannot be considered. Moreover, in a realistic building setting, there will be more possibilities for transportation of hot gases/smoke to different floors and compartments (e.g., through holes for pipes and conduits or even construction tolerances). These are also not considered in existing assembly-level tests.
- Combined testing of subsidiary systems, such as doors, compartment walls, structural fire protection, through-penetration firestops, and sprinkler system, is possible.

6.3 Limitations of Current Methods

- The main limitation of such a test facility is the high cost of experimentation as the building required overhauling of the fire protection system and relaying of the instrumentation wires before every experiment. If cost is not a concern, one may consider using modern wireless sensors and more elaborate protection for internal sensors (e.g., fire-resistant glass enclosure for video cameras) which may reduce consumption of certain sensors during the experiments.
- While the fire building has been designed and successfully used to simulate real fire scenarios, this attribute is a limiting factor as well. An experiment, once started, cannot be controlled and modulated. If a sensor cable burns during the experiment, it cannot be made good and loss of data must be endured. This can be countered to some extent by including redundancy in measurement systems, e.g., measurement of temperatures from inside the building through thermocouples and from outside the building through thermal imaging cameras.
- Due to the high cost and large number of interacting components, the number of scenarios that can be tested is limited.
- Since real fire scenarios are developed, each fire is different and unique from the other. Thus, such experiments cannot be used to provide standard ratings to different systems. However, they serve well for validation of tested systems in real scenarios.
- Instrumentation is very challenging due to the large distances within the building.
- As the experimentation requires standard civil/electrical work in addition to the façade installation, typically carried out by different agencies, it is difficult to control the variability. While it represents a realistic construction scenario, thereby leading to an almost real building fire (where multiple contractors might have worked), it creates challenges in the experimental regime.

6.4 Future Directions

It is clear that performance of different components of a façade (or a building) depends on a large number of factors and fire scenarios can be highly variable in nature. Thus, extending performance from standardized laboratory test to real fire scenarios is challenging. Even within laboratory testing, links between different scales of testing are important and require better understanding. Detailed computer models can help bridge these gaps in understanding but are currently at nascent stages of development due to the significant complexities that arise in fire modeling, especially with combustible façade systems. The authors believe that a combination of detailed computer models and standardized laboratory testing regimes may offer sufficient confidence in the fire performance of such systems. However, facilities like the one discussed in this work are essential to provide the necessary reality checks for laboratory testing methods and validation data for detailed computer models.

Bibliography

Abrantes V, Rangel B, Faria JMA (2017) The pre-fabrication of building facades. Springer, Berlin

AIAA (2011) Guide for the verification and validation of computational fluid dyamics simulations. American Institute of Aeronautics and Astronautics, 12700 Sunrise valley Drive, Suite 200, Reston

Aksamija A, Will P (2013) Sustainable facades: design methods for high-performance building envelops. Wiley, Hoboken

Anderson J, Bostrom L, McNamee RJ, Milovanovic B (2017) Modelling of fire exposure in facade fire testing. Fire Mater 42(5), 475–483

ASTM E1105-15 (2015) Standard test method for field determination of water penetration of installed exterior windows, skylights, doors, and curtain walls, by uniform or cyclic air pressure difference. ASTM International, West Conshohocken

ASTM E1300 (2016) Standard practice for determining load resistance of glass in buildings. ASTM International, West Conshohocken

ASTM E2268-04 (2016) Standard test method for water penetration of exterior windows, skylights, and doors by rapid pulsed air pressure difference. ASTM International, West Conshohocken

ASTM E283, E283M-19 (2017) Standard test method for fire tests of penetration firestop systems. ASTM International, West Conshohocken

ASTM E283, E283M-19 (2019) Standard test method for determining rate of air leakage through exterior windows, skylights, curtain walls, and doors under specified pressure differences across the specimen. ASTM International, West Conshohocken

ASTM E331-00 (2016) Standard test method for water penetration of exterior windows, skylights, doors, and curtain walls by uniform static air pressure difference. ASTM International, West Conshohocken

ASTM E547-00 (2016) Standard test method for water penetration of exterior windows, skylights, doors, and curtain walls by cyclic static air pressure difference. ASTM International, West Conshohocken

ASTM E783-02 (2018) Standard test method for field measurement of air leakage through installed exterior windows and doors. ASTM International, West Conshohocken

ASTM E966-18a (2018) Standard guide for field measurements of airborne sound insulation of building facades and facade elements. ASTM International, West Conshohocken

G. Srivastava and P. D. Gandhi, *Performance of Combustible Façade Systems Used in Green Building Technologies Under Fire*, Springer Transactions in Civil and Environmental Engineering, https://doi.org/10.1007/978-981-16-3112-2

Beyler CL (1986) Fire plumes and ceiling jets. Fire Saf J 11(1–2), 53–75

Blake P, Phylaktou H, Andrews G (2018) Validating FDS against a full-scale fire test. Technical report, University of Leeds, West Yorkshire

Bonner M, Rein G (2020) List of facade fires 1990–2020

Brito MC, Freitas S, Guimaraes S, Catita C, Redweik P (2017) The importance of facades for the solar PV potential of a mediterranean city using LiDAR data. Renewe Energy 111:85–94

Brookhaven National Laboratory (1976) An assessment of thermal insulation materials and systems for building applications. Technical Report 061-000-00094-1, Brookhaven National Laboratory, Washington DC, USA

BS 8118 (1991) Structural use of aluminum. Code of practice for design, BSI, UK

Development of a European approach to assess the fire performance of facades, June 2018

Donaldson B (1991) Exterior wall systems: glass and concrete technology, design, and construction. Technical Report STP 1034, 01-010340-10, ASTM, PA, USA

Dréan V, Schillinger R, Auguin G (2008) Fire exposed facades: numerical modeling of the LEPIR-2 testing facility. In: 2nd international seminar for fire safety of facades, Lund, Sweden

Dréan V, Girardin B, Guillaume E, Fateh T (2019) Numerical simulation of the fire behaviour of facade equipped with aluminium composite material-based claddings–model validation at intermediate scale. Fire Mater 43:839–856

Eurocode 3 Part 1-1 (2005) Eurocode 3: design of steel structures–Part 1–1: general rules and rules for buildings. European Standard, Management Centre, Brussels

Eurocode 3 Part 1-2 (2005) Eurocode 3: design of steel structures–Part 1–2: general rules–structural fire design. European Standard, Management Centre, Brussels

Eurocode 9 Part 1-1 (2005) Eurocode 9: design of aluminum structures–Part 1–1: general rules and rules for buildings. European Standard, Management Centre, Brussels

Eurocode 9 Part 1-2 (2005) Eurocode 9: design of aluminum structures–Part 1–2: general rules–structural fire design. European Standard, Management Centre, Brussels

Gandhi P, Jagdish V, Karthikeyan G, Chakravarthy A, Nakrani D, Ghoroi C, Srivastava G (2017) Performance of glass-ACP facade system in a full-scale real fire test in a G+2 structure. Procedia Eng 210:512–519

Giraldo MP, Avellaneda J, Lacasta AM, Rodriguez V (2012) Computer-simulation research on buildig-facade geometry for fire spread control in buildings with wood claddings. In: World conference on Timber Engineering, Auckland

Grenfell Tower Inquiry (2019) Grenfell tower inquiry, presented to Parliament pursuant to section 26 of the inquiries act 2005, ordered by the House of Commons

Guillaume E, Fateh T, Schillinger R, Chiva R, Ukleja S (2018) Study of fire behaviour of facade mock-ups equipped with aluminium composite material-based claddings, using intermediate-scale test method. Fire Mater 42:561–577

Hasemi Y (1984) Experimental wall flame heat transfer analysis for the analysis of upward flame spread. Fire Sci Technol 4(2), 75–90

Herzog T, Krippner R, Lang W (2004) Facade construction manual, 1st edn. Birkhauser-Publishers for Architecture, Basel, Boston. 978-3-7643-7109-8 978-3-7643-7031-2

Herzog T, Krippner R, Lang W (2018) Facade construction manual, 2nd edn. DETAIL Construction Manuals

Heskestad G (1983) Virtual origin of fire plumes. Fire Saf J 5:109–114

Heskestad G (1986) Fire plume air entrainment according to two competing assumptions. The proceedings of symposium (international) on combustion, pp 111–120

Heywood MD (2021) Best practice for the specification and installation of metal cladding and secondary steelwork. Technical report

Hidalgo JP, Torero JL, Welch S (2017) Experimental characterization of the fire behavior of thermal insulation materials for a performance-based design methodology. Fire Technology 53:1201–1232

Hu LH, Lu KH, Tang F, Delichatsios MA, He LH (2013) Effects of side walls on facade flame entrainment and flame height from opening in compartment fires. 03001, 2013. MATEC Web of conferences, 46, 05009
Huggett C (1980) Rate of heat release—implications for engineering decisions. In: Engineering applications of fire technology workshop, Gaithersburg, National Bureau of Standards
Hurley MJ, Gottuk DT, Hall JR Jr, Harada K, Kuligowski ED, Puchovsky M, Torero JL, Watts JM Jr, Wieczorek CJ (2016) SFPE handbook of fire protection engineering, 2nd edn. Springer, New York
ISO 16730-1 (2015) Fire safety engineering—procedures and requirements for verification and validation of calculation methods—Part 1: general. International Standards Organization
JIS A-1310 (2015) Test method for fire propagation over building facades
Klopovic S, Turan OGF (2001) A comprehensive study of externally venting flames - part II: Plume envelope and centre-line temperature comparisons, secondary fires, wind effects, and smoke management system. Fire Safety Journal 36:135–172
Kralj A, Drev M, Žnidaršič M, Černe B, Hafner J, Jelle B (2019) Investigations of 6-pane glazing: properties and possibilities. Energy Build 190:61–68
Lee Y-P, Delichatsios MA, Silcock GHW (2007) Heat fluxes and flame heights in facades from fires in enclosures of varying geometry. Proc Combust Inst 31:2521–2528
Lee Y-P, Delichatsios MA, Silcock GHW (2008) Heat flux distribution and flame shapes on the inert facade. In: Fire safety science—Proceedings of the ninth international symposium, pp 193–204
Lee Y-P, Delichatsios MA, Ohmiya Y (2012) The physics of the outflow from the opening of an enclosure fire and re-examination of Yokoi's correlation. Fire Saf J 49:82–88
Lee YP, Delichatsios MA, Ohmiya Y, Wakatsuki K, Yanagisawa A, Goto D (2009) Heat fluxes on opposite building wall by flames emerging from an enclosure. Proc Combust Inst 32(2), 2551–2558
Lewis MD (1995) Modern stone cladding: design and installation of exterior dimension stone systems. Technical Report Manual 21, 28-021095-10, ASTM, PA, USA
Lopez CR (2009) Comparison of wind-driven rain test methods for resident fenestration. Master's thesis, University of Florida, USA
MacDonald M (2004) Alternatives to glass mineral wool. Technical report, Competition Commission, Victory House, Brighton, BN1 4FY, UK
McGrattan K, McDermott R, Hostikka S, Floyd J (2014a) Fire dynamics simulator (Version 6)–User's guide. NIST Special Publication, NIST, USA
McGrattan K, Peacock R, Ooverholt AK (2014b) Fire model validation—eight lesssons learnt. In: Fire safety science—Proceedings of the eleventh international symposium, pp 958–968
McKenna ST, Jones N, Peck G, Dickens K, Pawelec W, Oradei S, Harris S, Stec AA, Hull TR (2019) Fire behaviour of modern facade materials–understanding the Grenfell Tower fire. J Hazardous Mater 368:115–123
Merritt FS, Ricketts JT (eds) (2001) Building design and construction handbook, 6th edn. McGraw-Hill, New York. 978-0-07-041999-5
Morgado HJL, Rodrigues JPC (2015) Balcony effect on the external fire spread into upper floors. J Fire Protect Eng 6(4), 255–273
Morton BR, Taylor GI, Turner JS (1956) Turbulent gravitational convection from maintained and instantaneous sources. Proc R Soc Lond Ser A Math Phys Sci 234(1196), 1–23
Nakrani D, Wani T, Srivastava G (2020) Simulation of pyrolysis and combustion of pine wood using two-step reaction scheme. In: Proceedings of SiF 2020—11th international conference on structures in fire. Brisbane, Australia. The University of Queensland
NBC (2016) National Building Code of India, SP 7, vol 1. Bureau of Indian Standards, New Delhi, p 2016
Nelson HE (1988) An engineering view of the fire of May 4, 1988 in the first interstate bank building, Los Angeles, California. Technical Report NISIR-89-4061, US Department of Commerce, National Institute of Standards and Technology, National Engineering Laboratory, Center for Fire Research, Gaithersburg, MD, 20899

Nelson PE, Kroll RE (1996) Exterior insulation finish systems (EIFS): materials, properties, and performance. Technical Report E-6, ASTM, PA, USA

Nishio Y, Noguchi T, Yoshioka H, Kanematsu M (2018) A study on heat output for JIS A 1310 test method for fire proagation over building facades. J Environ Eng, AIJ 83(753):851–860

Oleszkiewicz I (1989) Heat trnasfer from a window fire plume to a building facade. Technical Report HTD- 123, American Society of Mechanical Engineers, USA

Oleszkiewicz I (1990) Fire exposure to exterior walls and flame spread to combustible cladding. Fire Technol 26:357–375

Oleszkiewicz I (1991) Vertical separation of windows using spandrel walls and horizontal projections. Fire Technol 27:334–340

Pagni PJ, Joshi AA (1991) Glass breaking in fires. In: Fire safety science—Proceedings of the third symposium, pp 791–812

Patterson M (2011) Structural glass facades and enclosures. Wiley, Canada

Peng L, Ni Z (2016) Experimental study of window-ejected flame and plume on glass curtain walls. In: 2nd international seminar for fire safety of facades, Lund, Sweden. MATEC Web of Conferences, 46, 05009

Reid RN (2000) Roofing and cladding systems: a guide for facility managers. Fairmont Press, USA

Routley JG (1988) Interstate bank building fire. Technical Report USFA-TR-022, US Fire Administration, Los Angeles, California, USA

SFPE G.06 (2011) Engineering guide: guidelines for substantiating a fire model for a given application. Society for Fire Protection Engineers, USA

Srivastava G, Ghoroi C, Gandhi P, Jagdish V, Karthikeyan G, Chakravarthy A, Nakrani D (2018) Development of a unique full-scale real-fire facade testing facility at IIT Gandhinagar. Current Sci 115(9), 1782–1787

Srivastava G, Nakrani D, Ghoroi C (2020) performance of combustible facade systems with glass, ACP and firestops in full-scale, real-fire experiments. Fire Technol 56(4), 1575–1598

Strother EF, Turner WC (1990) Thermal insulation building guide. Krieger Publishing Company, USA

Summers PT, Chen Y, Rippe CM, Allen B, Mourtiz AP, Case SW, Lattimer BY (2015) Overview of aluminum alloy mechanical properties during and after fire. Fire Sci Rev 4(3)

Tang F, Hu LH, Delichatsios MA, Lu KH, Zhu W (2012) Experimental study on flame height and temperature profile of buoyant window an under-ventilated compartment fire. Int J Heat Mass Transf 55:93–101

Thomas PH (1981) Testing products and materials for their contribution to flashover in rooms. Fire Mater 5(3), 103–111

Tittow WV (1986) PVC Technology, 4th edn. Elsevier Applied Science Publishers, London

Unguresan PV, Porumb RA, Petreus D, Pocola AG, Pop OG, Balan MC (2017) Orientation of facades for active solar energy applications in different climatic conditions. ASCE J Energy Eng 143(6):04017059

University of Queensland (2021) Cladding materials library

Vanhoutteghem L, Skarning GCJ, Hviid CA, Svendsen S (2015) Impact of façade window design on energy, daylighting and thermal comfort in nearly zero-energy houses. Energy Build 102:149–156

Walton WD, Thomas PH (2016) Estimating temperatures in compartment fires. In: SFPE handbook of fire protection engineering, 2

White N, Delichatsios M (2015) Fire hazards of exterior wall assemblies containing combustible components. Springer, London

Wickström U (2008) Adiabatic surface temperature and the plate thermometer for calculating heat transfer and controlling fire resistance furnaces. In: Fire safety science—Proceedings of the ninth international symposium, pp 1227–1238

Wickström U (2016) Temperature calculation in fire safety engineering. Springer, Berlin

Yanagisaw A, Jo A, Nakao T, Wakatsuki K, Ohmiya Y (2007) Fire plume ejected from an opening in unconfined space, part 2: generation limit of external flame. Fire Sci Technol 26(4), 505–510

Yokoi S (1960) Study on the prevention of fire spread caused by upward hot current. Technical Report 34, Building Research Institute, Japan

Nishio Y, Yoshioka H, Noguchi T, Kanematsu M, Ando T, Hase Y (2016) Fire propagation over combustible exterior facades exposed to intensified flame in Japan. In: 2nd international seminar for fire safety of facades. Lund, Sweden. MATEC web of conferences, 46, 05009

Zhao G, Beij T, Merci B (2016) Application of FDS to under-ventilated enclosure fires with external flaming. Fire Technol 52:2117–2142

Zukoski E, Kubota T, Cetegen B (1981) Entrainment in fire plumes. Fire Saf J 3:107–122

Index

G. Srivastava and P. D. Gandhi, *Performance of Combustible Façade Systems Used in Green Building Technologies Under Fire*, Springer Transactions in Civil and Environmental Engineering, https://doi.org/10.1007/978-981-16-3112-2